T0190155

Historical Geography and Geosciences

This book series serves as a broad platform for contributions in the field of Historical Geography and related Geoscience areas. The series welcomes proposals on the history and dynamics of place and space and their influence on past, present and future geographies including historical GIS, cartography and mapping, climatology, climate history, meteorology and atmospheric sciences, environmental geography, hydrology, geology, oceanography, water management, instrumentation, geographical traditions, historical geography of urban areas, settlements and landscapes, historical regional studies, history of geography and historic geographers and geoscientists among other topically related areas and other interdisciplinary approaches. Contributions on past (extreme) weather events or natural disasters including regional and global reanalysis studies also fit into the series.

Publishing a broad portfolio of peer-reviewed scientific books Historical Geography and Geosciences contains research monographs, edited volumes, advanced and undergraduate level textbooks, as well as conference proceedings. This series appeals to scientists, practitioners and students in the fields of geography and history as well as related disciplines, with exceptional titles that are attractive to a popular science audience.

If you are interested in contributing to this book series, please contact the Publisher.

More information about this series at http://www.springer.com/series/15611

Drielli Peyerl

The Oil of Brazil

Exploration, Technical Capacity, and Geosciences Teaching (1864–1968)

Drielli Peyerl
Institute of Energy and Environment
University of São Paulo
São Paulo, Brazil

ISSN 2520-1379 ISSN 2520-1387 (electronic)
Historical Geography and Geosciences
ISBN 978-3-030-13886-8 ISBN 978-3-030-13884-4 (eBook)
https://doi.org/10.1007/978-3-030-13884-4

Library of Congress Control Number: 2019932674

Translation from the Portuguese language edition: O Petróleo no Brasil by Drielli Peyerl,

This Springer imprint is published by the registered company Springer Nature Switzerland AG
The registered company address is: Gewerbestrasse 11, 6330 Cham, Switzerland

To Jeferson Peyerl

To Frederico Waldemar Lange (1911–1988) for everything he afforded, without even knowing, to my life.

Foreword

The period from 1854 to 1968 was particularly emblematic for the oil exploitation in Brazil. The word petroleum, mentioned for the first time in the decree published on November 1864, started to have an essential role in the country economy in the following decades, when exploratory activities to the search of the "black gold" had begun. In the following decades, decrees had been published, private foreign companies had been established, drilling had been made, and multiple activities for the technical qualification had been developed, which have led to a nationalist politics that culminated in the creation of the National Petroleum Council (Conselho Nacional do Petróleo) and the discovery of the first oil well in the district of Lobato, Salvador City, Bahia State, in 1939. From that moment onwards, a new phase started in the economic politics related to the exploration and petroleum industry, progressing in stride with the creation of Petrobras, in 1953, and its Petroleum Improvement and Research Center and, approximately 13 years later, in 1966, the beginning of the Research and Development Center Leopoldo Américo Miguez de Mello (CENPES), the first one destined to the improvement and specialization of manpower, and the second center, created after the first one was extinguished, as a center in excellency on scientific and technological research for the petroleum industry in Brazil. I was a long path that for over a century, relied on many comings and goings, from the politic and economic point of view. This trajectory of oil in Brazil needed a synthetic approach to the events and main characters related to the history of oil research in Brazil, and it is precisely with a concise and objective approach that the text of the historian Drielli Peyerl rises, written with unmatched mastery.

Holding a Bachelor's degree in History, with Geography Degree, Master's Degree in Geography from the State University of Ponta Grossa, in Paraná, and Doctor in Sciences from the University of Campinas, in São Paulo, Drielli focuses on the study of the formation of the petroleum industry in its more intriguing period, in addition to approaching its relation and importance to the education of geosciences in the country.

Drielli started her work a starting point the private documentation of one of the great icons of the petroleum research in the country, the paleontologist Frederico Waldemar Lange, which, added to a wide documentary research in the Petrobras archives and other institutions, including abroad, allowed the elaboration of an instigating text about the several paths followed by the Brazilian petroleum industry.

Aiming at the good understanding of the reader, Drielli starts her piece with an introductory text, in which she lists concisely the main historical aspects that are to be observed by the reader and talks about the same in the three following chapters: in the first, the author approaches the aspects of the initiatives and research, both private and governmental, that had led to the discovery of oil and the creation of the National Petroleum Council, involving all the events between 1938 and 1961; in the second, she talks about the participation and the role of Brazilian and foreign technicians in the formation of the petroleum industry; in the third and last, she focuses on the formation of the improvement and research centers, as well as weaving considerations about the importance of professionalizing courses for the technician staff of Petrobras.

Using unpublished documents, a great bibliography and instigating information, Drielli manages to trace the efforts and initiatives that had led to the enrichment and the improvement of the Brazilian petroleum industry in a work that, without a doubt, will become reference in the historical research in the petroleum and its influence in the education of the geosciences in Brazil.

Rio de Janeiro, Brazil
July 2016

Antonio Carlos S. Fernandes

Acknowledgements

To Prof. Dr. Silvia Fernanda de Mendonça Figueirôa, advisor, friend, and one of the most brilliant and intelligent people that I have had the privilege to meet, who also gave me the opportunity to fulfill my goals and dreams.

To Prof. Dr. Elvio Pinto Bosetti, advisor, friend, and the one who believed in my work, who also taught me to work as a team and showed me that this is actually possible in the academy.

To Prof. Dr. Brian Frehner, friend and the person responsible for introducing me to the theme History of Energy and History of Environment in my academic life.

To São Paulo Research Foundation (Fundação de Amparo à Pesquisa do Estado de São Paulo—FAPESP) for the grant awarded during Ph.D. (Process No. 2010/14857-2), Post-Doctorate (Process No. 2014/06843-2 and 2015/03244-3) and Young Investigator (Process No. 2017/18208-8 and No. 2018/26388-9).

To my parents, Irvando Luis Peyerl and Neusa do Carmo Hacke, and to my brother, Jefferson Peyerl, for the unconditional support in all moments.

In special, to Amy Randolph, Carlos Roberto dos Anjos Candeiro, Cristina de Campos, Dominique Mouette, Edmilson Moutinho dos Santos, Elisamara Aoki Gonçalves (Translator), Evandro Mateus Moretto, Julio R. Meneghini, Karen Louise Mascarenhas, Matthew R. Silverman, Nathália Weber Neiva Masulino, Rafael Alfena Zago, Robert H. Dott Jr. (1929–2018), Raquel Rocha Borges, Tyler Priest, Walter Oscar Serrate Cuellar, and William Brice.

To the member of my Ph.D.'s work, Profs. Dr. André Tosi Furtado, Dr. Antônio Carlos S. Fernandes, Dr. Jefferson de Lima Picanço, and Dr. Maria Amelia Mascarenhas Dantes.

To institutions, Linda Hall Library (staff), Oklahoma State University, Research Centre for Gas Innovation, University of Missouri-Kansas City, State University of Ponta Grossa, University of Campinas, and University of São Paulo (Institute of Energy and Environment), who contributed in many ways to my formation and for the development of my research.

To the Geosciences Institute (University of Campinas), to the professors and servers, to the Post-Graduation Program in Teaching History of Earth Sciences (PEHCT/University of Campinas), to the State University of Campinas, to the Library of the Institute of Philosophy and Human Sciences (IFCH/University of Campinas), to the Archive Frederico Waldemar Lange (1911–1988) (State University of Ponta Grossa), and to the section of the rare works collection from the Library CENPES/Petrobras.

To the Palaios Group (State University of Ponta Grossa/National Council for Scientific and Technological Development).

My sincere thanks to Dr. Michael Leuchner for the opportunity to publish this book.

My thanks to UFABC publisher for the Portuguese version of this book published in 2017.

At last, although there is a lot to be said, I thank my friends and colleagues who contributed to this process and whose wise advices and support are acknowledged and very much appreciated.

Contents

Abbreviations

CAGE	Campaign for Geologists Formation (Campanha de Formação de Geólogos)
CAPES	Coordination of the University Education Improvement Personnel (Coordenação de Aperfeiçoamento de Pessoal de Nível Superior)
CENAP	Center for Improvement and Petroleum Research (Centro de Aperfeiçoamento e Pesquisas de Petróleo)
CENPES	Research and Development Center Leopoldo Américo Miguez de Mello (Centro de Pesquisas e Desenvolvimento Leopoldo Américo Miguez de Mello)
CGB	Brazilian Geological Commission (Comissão Geológica do Brasil)
CGG	Geographical and Geological Commission of São Paulo (Comissão Geográfica e Geológica de São Paulo)
CNP	National Petroleum Council (Conselho Nacional do Petróleo)
CPDOC	Center of Research and Brazilian Contemporary Historical Documentation (Centro de Pesquisas e Documentação de História Contemporânea do Brasil)
DEPEX	Exploration Department from Petrobras (Departamento de Exploração da Petrobras)
DEXPRO	Department of Exploration and Production from Petrobras (Departamento e Exploração e Produção da Petrobras)
DIVEX	Exploration Division Surface Sector (Setor de Superfície da Divisão de Exploração)
DNPM	National Department of Mineral Production (Departamento Nacional de Produção Mineral)
EMOP	School of Mines of Ouro Preto (Escola de Minas de Ouro Preto)
FNFI	National Faculty of Philosophy (Faculdade Nacional de Filosofia)
FRONAPE	National Tanker Fleet (Frota Nacional de Petroleiros)
IPT	Technological Research Institute (Instituto de Pesquisas Tecnológicas)

ITA	Technological Aeronautics Institute (Instituto Tecnológico da Aeronáutica)
PED	Strategic Development Program (Programa Estratégico de Desenvolvimento)
Petrobras	Petróleo Brasileiro S.A.
PIPMOI	Intensive Industrial Labor Preparation Program (Programa Intensivo de Preparação de Mão de Obra Industrial)
SBP	Brazilian Society of Paleontology (Sociedade Brasileira de Paleontologia)
SENAI	National Industrial Learning Service (Serviço Nacional de Aprendizagem Industrial)
SFPM	Mineral Production Promotion Survey (Serviço de Fomento da Produção Mineral)
SGMB	Geological and Mineralogical Survey of Brazil (Serviço Geológico e Mineralógico do Brasil)
SSAT	Technical Improvement Supervision Sector (Setor de Supervisão do Aperfeiçoamento Técnico)
UB	University of Brazil
UCLA	University of California
UEPG	State University of Ponta Grossa
UFBA	Federal University of Bahia
UFRGS	Federal University of Rio Grande do Sul
UFRJ	Federal University of Rio de Janeiro
UNICAMP	University of Campinas
URGS	University of Rio Grande do Sul
USP	University of São Paulo

List of Figures

List of Tables

List of Graphs

List of Maps

List of Organizational Charts

Presentation

In history, the establishment of secular clippings is always subject to decisions—in the background, choices—supported by available data and questions with which we question reality. In that way, when did the work, which gave rise to this book start?

I risk saying that it was born in an elevator in Curitiba (Paraná State, Brazil), when a fortuitous meeting during the Brazilian Congress of Geology of 2008 brought up the partnership between Drielli Peyerl, and me. From then on, I knew, and I followed the Master Degree research that she developed with the State University of Ponta Grossa, under the orientation of Prof. Dr. Elvio Pinto Bosetti. I had the satisfaction to receive her as a special student in the University of Campinas (Unicamp) in 2009, where she deepened her knowledge in the History of the Geosciences. I had a greater satisfaction when I had the chance to be part of her board of Master Degree defense and when I became her Ph.D. advisor in 2010.

The text here published is her thesis, defended in 2014 throughout the 4 years of her Ph.D., supported for by the FAPESP, I was a witness of her seriousness and restless disposal for serious, deep and rigorous research, primary documentation, old or more recent sources and several bibliographies, all that Drielli managed to compile (in Brazil and abroad), process, compare and organize, producing a fluent narrative and innovative perspective. The biggest merits (among so many) of this book are the richness of sources—among which stands out the file of the paleontologist Frederico Waldemar Lange, figurehead in the early stages of Petrobras—and the theme itself. After all, much has been written about petroleum in Brazil—without running out the theme—and much more will be written if we want to understand this story. However, this book focuses on the Petrobras from the perspective of scientific–technical human resources formation that the company had to undertake, at different levels of complexity, pari passu its own institutional construction. Then, the challenge of geologically map the territory emerged, aiming to locate possible fields of exploration, to extract, to process and to distribute the black gold in a country with a history of formation of human resources in higher education had only approximately 150 years, if we consider the initial landmark as being the two Schools of Medicine established in 1808 (in Bahia and Rio de Janeiro States) and the Real Military Academy, founded in 1810. The narrative of this context mixes, forcibly, the economy, the politics, sciences, technologies, the society, and the international relations, interlaced in the assembly of a scenario that, from projects and program of courses, contributes to Brazil thinking in the effervescent decades of the first half of twentieth century.

Undoubtedly, this is a rich reading for academics and laypeople that want to understand Petrobras and the petroleum under a new and original point of view.

Campinas, Brazil
July 2016

Silvia F. de M. Figueirôa

1 Introduction

In 1864, Brazil was in the Imperial Period (1822–1889), more precisely in the Second Reign (1840–1889), which was stricken by important facts, such as the beginning of the Paraguayan War (1864–1870) and the issue of the decree No. 3.352-A, from 30 November 1864, when for the first time, the word *petroleum* is mentioned in the Brazilian Legislation.

This decree will be the starting point of our discussion, in which we will initially approach the exploratory activities involving the petroleum in Brazil, enabled by other 73 federal decrees issued between 1864 and 1938—i.e., until the creation of the National Petroleum Council—which authorized the enterprise of natural person and legal person (foreigner and national) in those activities in different sites from the national territory.

Scientific and geological events also marked the period mentioned, with the creation of the Brazilian Geological Commission and Ouro Preto School of Mines, in 1875, the Geographical and Geological Survey of São Paulo, in 1886, the Commission for Studies on Coal Mines, in 1904, and the Geological and Mineralogical Survey of Brazil, in 1907.

In 1897, the official presence of private enterprise, by the means of foreigner companies, was consolidated mainly by the North American industry operation, which contributed to significant changes in the research and development scenario of petroleum exploratory techniques in Brazil. On that same year, the deep drilling also takes place, an action that leads to technical and empirical development in the search of petroleum. This period is marked by the technical capacitation through manuals and help from foreigners that know about the technique. In 1913, Mexico was another country that obtained an authorization to operate in Brazil, with The Anglo Mexican Petroleum Products company, founded by the British Engineer Weetman Dickinson Pearson (1856–1927), who had already started a policy to market products derived from petroleum also the obtainment of the petroleum refining control in Latin American countries.

From 1930, political and economic changes modified the rhythm of the petroleum research and exploration in Brazil, out of those, a nationalist policy is noted, especially when the Mining Code (1934) or ordained, creating the National Petroleum Council (1938) and when the first petroleum well is discovered in the district of Lobato, Salvador City, Bahia State (1939).

Companies and foreign professional entities, such as federal agencies, have contributed to the geology study of the area and to perform drillings in several points of the Brazilian territory. Although, a few difficulties were directly related to the purchase of materials, high costs of importation—usually from the United States—and the

D. Peyerl, *The Oil of Brazil*, Historical Geography and Geosciences,
https://doi.org/10.1007/978-3-030-13884-4_1

lack of Brazilian professionals, which demanded the need to foreign aid.

With the National Petroleum Council, a new political and economic phase arises, aiming at the petroleum exploration and industry. It was nationalized before its discovery, in a phase characterized by external and internal conflicts between national and great petroleum international groups regarding its exploration and refining. In such dispute between nationalists and *entreguistas* (those who defend opening to external capital), there is an emphasis on the state petroleum monopoly, which culminated, from 1947, in the Campaign named "The petroleum is ours" (*O petróleo é nosso*).

One of the biggest problems faced by the National Petroleum Council was the lack of qualified manpower in the national territory for the petroleum refining and exploration activities. One of the first alternatives given by the National Petroleum Council was the formalization of agreements with foreign companies, contracted by the agency to set up here to train and Brazilian manpower in their line of work. Alongside such training, a few Brazilian professionals were sent abroad for professional improvement and specialization. Nevertheless, those attempts were not solving the problem of lack of professionals, whether by the number of Brazilian people that was being trained or by the reduced number on manpower that remained here, as a number of Brazilian professionals were sent abroad. The refining industry was growing together with the continuous search for new oil wells, that manner, professionals such as geologists, geophysicists, oil engineers, maintenance engineers, drillers, blade operators, mechanics, designers, seismograph operators, among others were increasingly necessary.

In 1952, the National Petroleum Council invested in the creation of a sector that was able to form specialized professionals, the Technical Improvement Supervision Sector (SSAT), which aimed to generate technical specialized manpower and use it as an instrument for operation. Such an initiative would carry out deep changes in the directions of the geosciences teaching in Brazil. On the same year, the SSAT created the first petroleum refining course, together with the University of Brazil.

This action from the National Petroleum Council was mainly related to the need for the country to develop its own know-how and to stop relying on knowledge and techniques from other countries. That way, the country had reached a moment when there has been an investment in the development of techniques and equipment, mainly in the formation of petroleum-related professionals.

In 1953, Petrobras was created, a company with mixed economy, which has just gradually absorbed the activities from the National Petroleum Council and took responsibility for the state petroleum monopoly. With these activities being absorbed, Petrobras also assimilated the problems of the National Petroleum Council, mainly centered on the lack of specialized manpower and the permanent presence of foreigners in the petroleum exploration and industry. That occurred after several nationalist attitudes, like the previously mentioned campaign "The petroleum is ours", which was slowed down by the creation of the Petrobras.

After absorbing the SSAT, in 1955, the company invested in the creation of the Center for Improvement and Petroleum Research (CENAP), aiming to promote courses for the improvement and specialization of the manpower and to implement technological exploratory petroleum-related research. The courses were destined to high school, technical degree, and graduation. The offer of grad-level improvement and specializations courses provided a higher contact with the university teaching in Brazil.

The improvement and specialization courses applied by CENAP corresponded to the Petroleum Refining, Petroleum Equipment Maintenance, Introduction to Geology, Petroleum Geology, and Petroleum Engineering courses. Courses in the Geology field were the highlight, which contributed to the opening of the Geology course in the country, in 1957. From that moment, Petrobras is consolidated, not only as one of the biggest companies in the oil industry but also as an institution that provides geosciences teaching in Brazil.

In 1966, the CENAP was extinguished, and the activities of the Research and Development Center Leopoldo Américo Miguez de Mello (CENPES) started, which aimed to perform studies with scientific and/or technological interest to the oil industry—different purpose as the one proposed by the CENAP. The improvement and specialization courses were continuously be offered in the company, according to Petrobras needs for specific formations in the petroleum research. In this process, Petrobras became the flagship of the economy and scientific, technological, and innovative research in the country when it comes to geosciences.

However, how were those courses structured? What was the objective of the National Petroleum Council and Petrobras when they formed their own manpower? How did it influence in the creation of the first Geology courses in the country in 1957? Who would teach the disciplines?

Since the creation of the National Petroleum Council to the work made by Petrobras, the initial objective with the creation of the courses was temporary, which was forming only the necessary number of professionals and gradually transferring the courses to the Brazilian Universities. That would be possible as the courses were structured based on the descriptions of graduation courses and tied, through agreements, to Brazilian and foreign universities.

What seems to be a very simple solution (transformation of the courses offered by Petrobras in graduation or specialization courses) occurred partially, as the growth of the oil industry demanded other formations, and given the circumstances, it was not suitable to the moment to create a graduation course that specifically aimed at petroleum studies. That manner, Petrobras continued to invest in improvement and specialization courses, which culminated, decades later, in the creation of the Petrobras University.

For the structuring, elaboration, and application of the courses, the partnership with different teaching institutions, including foreign ones and with the participation of foreign professionals (acting like teachers and coordinators of the courses), was essential to the implementation of such courses and the establishment of the Brazilian know-how associated to the petroleum. The presence of foreigners in the country, working for the National Petroleum Council and for Petrobras, in a period marked by a strong nationalism, occurred in a very tense manner, so to speak. The idea transmitted by the National Petroleum Council and by Petrobras was to replace foreigners by Brazilian people; however, such process did not occur immediately, but it was made gradually, with the internal transformations of the country associated with the necessity at the moment.

That occurred due to the reasons that we will show in this book, and such changes were mainly made by the Exploration Department from Petrobras (DEPEX).

The Exploration Department was initially headed—from October 1954—by the North American geologist and an important character in the petroleum exploration in Brazil, Walter Karl Link (1902–1982). Link has complemented the department with an organizational structure in the molds of the North American industry and carried out previous studies of the Brazilian sedimentary basin, which resulted in the presentation of a detailed report comprising his 6 years of research in Brazil. The geologist highlighted that the research should target the continental platform or other countries, as the Brazilian territory was not rich in petroleum. Link was heavily criticized for those statements, and only years later, Petrobras would recognize that he was right.

On January 1, 1961, Link was replaced by the paleontologist, Frederico Waldemar Lange (1911–1988), the most capable professional to assume the role of Exploration Department Chief. In this period, internal changes started within the Petrobras, related to the department subdivisions.

The structure of the book follows a chronological order, which ends on 1968, when through the advances mainly by geophysics and offshore exploration techniques the first offshore well is discovered in the Guaricema field (Sergipe-Alagoas Basin) and oil exploration in the Campos Basin began in the same year. At that point,

Petrobras has invested excessively in the petroleum exploration research in the continental platform, including courses aiming at marine drilling techniques.

In this book, we will describe the elements that have contributed to the scientific and technical development of the petroleum exploratory research in Brazil. Among them, the exploratory points, qualification of techniques and improvements, and specialization of manpower by the National Petroleum, Council and mainly by Petrobras can be found, which directly rely on the foreign participation for the formation of own know-how.

We reinforce that it is not our intention—also not possible—to include all facts and acts that occurred during nineteenth and twentieth centuries regarding the petroleum, for that reason, we would not dare to propose a "real story"—term used and worked by Braudel (1992, 1993), Bloch (2001, 2009) and Febvre (1989). To do that, the reference used in this book helps and supports the construction and gathering of the scientific and technical development aspects related to the petroleum together with the selection of the sources used, based on the Science History.

In order to answer the overexposed inquiries and to describe the scientific and technical development process, the use of primary sources was essential to the unfolding of the book. In the Portuguese version, to keep the originality of the classics and old works/documents used, but in the English version, we chose to correct the orthography of that period. A specific case is that for a few times, the word Petrobras will appear with an accent: Petrobrás.[1]

The main sources used in this book belong to the private collection of the paleontologist Frederico Waldemar Lange (1911–1988). It is composed approximately by 120 boxes of files with much information related to the Petrobras and the studies about geology, paleontology, petroleum, and technical and scientific progresses that marked the period that Lange started his career, still young, as a self-taught paleontologist.[2] We can characterize the collection Frederico Waldemar Lange as a scientific research collection, with no nationalist characteristics. In his area of professional action, Lange is considered as a pioneer in the micropaleontology in Brazil—area that significantly contributed to the discovery of petroleum—being nationally and internationally recognized by his work.

Although this book does not focus on the Lange[3] path, we highlight that his private collection comprises several letters from scientists, geologists, and paleontologists, in addition to the internal Petrobras reports, newspaper clippings, photographs, and other materials that were used and served as base for this research.

In addition to the primary and secondary sources, we had access to the rare books collection sector from the CENPES/Petrobras library, which contributed to complement the information found in the Frederico Waldemar Lange collection, as well as the Philosophy and Human Sciences Institute Library from UNICAMP, whose collection has several books regarding the Petroleum History. Also, interviews made by the Center of Research and Brazilian Contemporary Historical Documentation (CPDOC) and me and also the documents available online were consulted. A document research was also made in different countries, like the United States, France, and Mexico.

As a building block of this book, we used two methods: the hermeneutical and net. At first, searched a method that allowed the dialogue between sciences—in the sense of interpreting and working with the sources presented in this research—the hermeneutical method and

[1] Originally Petrobrás, the name of the company was changed to Petrobras, although it is an oxytone that ends with an "a" (followed by an "s"), according to the Portuguese rule. n° 7.565 of 1971, in compliance with the Brazilian Academy of Letters (Academia Brasileira de Letras) and the Lisbon Sciences Academy (Academia das Ciências de Lisboa), which say that no acronym should take an accent in the Portuguese language.

[2] The consolidation of his career as a paleontologist also happened in the Parana Museum, and from 1955, he works in Petrobras, being one of the first paleontologists hired by the company.

[3] For more information, see Peyerl (2010).

interpretation means, comprehension and language that enabled to reveal the communication dimension inherent to the actions through which the man establishes a relation with what one himself builds (Bombassaro 1992)—in this case, the science itself.

The French philosopher Paul Ricouer (1913–2005) states that, within hermeneutics, one starts from the conflict finding and the interpretations, which might be seen in different ways, that is emphasized by the author in his piece of work *De l'interpretation. Essai sur Freud*, from 1965, which "there is no general hermeneutics, there is no universal canons to the exegesis, but separate and opposed theories treating the interpretation rules" (Dortier 2010: 269).

With the hermeneutical method, one searched to interpret the sources diversity of the book, allowing that such comprehension/interpretation brought the enlargement of our horizon of possibilities, thus highlighting that "writing a story about a period means to find assertions that could never be made in such period" (Alberti 1996: 52):

> It is fascinating to acknowledge that, as higher as our effort and grammar and history preparation, our comprehension of the other will never be full and finite. From that, it is possible to infer that the interpretations can be indefinitely rebuilt, with new angles and point of views, conditioned by the private position of each interpreter (Dortier 2010: 16).

Therefore, "the researcher, when working meticulously about this communicative material", with several interpretations and creations of meaning, "also become one-self one more interlocutor, interacting with the dialogue circuit of the knowledge production" (Carvalho 2003: 297). Thus, we can point out three examples of the hermeneutical method application in this book: first, the definition of the research site, as well as the type of sources to be worked; second, the precise screening of the sources according to the study; and third, the interpretation of the analyzed and used sources.

In order to complement the analyses achieved by the hermeneutical method, we adhered to the net method, which offered support to the comprehension of the relations established between human attitudes related to the technique development, that way associating the method in a private manner to the actor–network theory.[4] According to Wilkinson (2004), the actor–network theory, although used many times as a methodology, in practice it reached the status of a theory and it can be understood as:

> […] human and non-human bindings – which in turn, are more bindings too – setting, therefore, a network tangle that fragment any robustness in microconnections or disconnections. Such tangle allows us thinking not in terms of unit, but in a process dynamism and always constant of associations (Nobre and Ribeiro Pedro 2010: 48).

These constant associations and connections are capable of producing changes by the articulation of different elements, noticeable to the study of that time, using the primary sources mentioned.

That manner, the book was structured in three parts:

The first part of the book or the first part of the book chapter, entitled "The petroleum comes to light", begins in 1864, when the word petroleum comes to light for the first time in a federal decree, that manner it is described the pursuit for petroleum in the Brazilian territory together with the investment from private and governmental initiatives, which have contributed to the development of the technological petroleum research —many times oriented to the empirical and technology importation dependent method. The first book chapter ends in the years 1939, when the first sub-commercial oil well was discovered, and 1941, when the first continental commercial oil well is discovered in Brazil.

The second part, named "The know-how formation (1938–1961)" reveals the technical–scientific development since the creation of the National Petroleum Council (1938) and Petrobras

[4]Actor–network theory (ANT) was initially developed from the Studies of the Science and Technology (Edge 1994; Williams and Edge 1996; Button 1993; Grint and Woolgar 1997; Mackenzie and Wajcman 1999; Pinch and Bijker 1984, 1987), being the product of discussions from a group of French and English anthropologists, sociologists and engineers and associated, among them are Bruno Latour, Michale Callon and John Law. For more information, see: Alcadipani and Tureta (2009).

(1953), addressing the joint work between Brazilians and foreigners in the development of the research and exploration of petroleum in Brazil. It has a main analysis point, Petrobras Exploration department, seen through numbers and information about the constitution and working process until 1961.

On the third part, designated "Improvement, professionalization and geosciences teaching", an initial overview of the consolidation activities is suggested, made by the initiatives of the National Petroleum Council and Petrobras, with the creation of the sector of Technical Improvement Supervision Sector (SSAT), in 1952, the Center for Improvement and Petroleum Research (CENAP), in 1955, and the Center for Improvement and Research Leopoldo Américo Miguez de Mello (CENPES), in 1963—but activities started in 1966. With those initiatives, we have actions of technical and professional formation of Brazilian manpower implemented by the CNP and Petrobras, related to the investigation, study, and exploration of the oil. Together with those factors, there is still the contribution to the development of geosciences teaching in Brazil. This is how it ends this last part, in 1968, with the discovery of the first offshore well in the country.

References

Alberti V (1996) A existência na história: revelações e riscos da hermenêutica. Estudos históricos – Historiografia 9(17):31–57. Rio de Janeiro

Alcadipani R, Tureta C (2009) Teoria ator-rede e análise organizacional: contribuições e possibilidades de pesquisa no Brasil. Organizações & Sociedade 16 (51):647–664

Bloch M (2001) Apologia da história ou o ofício de historiador. In: Zahar J (ed). Rio de Janeiro

Bloch M (2009) A sociedade feudal. Edições, Coimbra, p 70

Bombassaro LC (1992) *As fronteiras da epistemologia*: uma introdução ao problema da racionalidade e da historicidade do conhecimento. Vozes, Petrópolis

Braudel F (1992) Reflexões sobre a história. Martins Editora, São Paulo

Braudel F (1993) O Mediterrâneo e o mundo mediterrâneo na época de Filipe II. Martins Fontes, São Paulo

Carvalho ICM (2003) Biografia, Identidade e Narrativa: Elementos para uma análise hermenêutica. Horizonte antropológico 9(19). Porto Alegre

Dortier JF (2010) Dicionário de Ciências Humanas. Martins Fontes, São Paulo

Febvre L (1989) Combates pela história. Editorial Presença, Lisboa

Nobre JC de A, Ribeiro Pedro RML (2010) Reflexões sobre possibilidades metodológicas da Teoria ator-rede. Cadernos Unifoa (14):47–56

Peyerl D (2010) A trajetória do paleontólogo Frederico Waldemar Lange (1911–1988) e a História das ciências. Thesis, State Ponta Grossa University, Ponta Grossa, 116 f

Wilkinson J (2004) Redes, convenções e economia política: de atrito à convivência. In: Encontro Anual da ANPOCS, 28. Caxambu. *Anais*. ANPOCS, Caxambu, pp 1–31

2 The Petroleum Comes to Light (1864–1941)

2.1 Initiatives and Technological Research of Petroleum

> Petroleum – It came up a Petroleum mass mixed with clay and sand, extracted from Taipú-mirim close to Barra de Camamú, less than 5 miles from the northwest of the Village and L. side of the river with the same name […]. With the aid of a drill it was reached, on November 1854, to 23 feet under the ground, always finding gravel with bituminous clay, also a vessel containing approximately three pounds of pure petroleum, extracted by distillation outdoor in piles or bins […]. By the stated this Petroleum is found, it deserves well the name they give, tar or pixe mineral […] (Petróleo 1855: 140).

Ten years after the referred fact in the above-mentioned extract, related to the oil extraction, taken from the important journal *O Auxiliador da Indústria Nacional*, we have a unique and official event that started and transformed the course of petroleum-related research in Brazil: the decree no. 3.352-A, from November 30, 1864, in which, for the first time, the word petroleum[1] is mentioned in the text of the Brazilian Legislation. Such decree had granted to the Englishman "Thomaz Denuy Sargent capability for a ten-year period, by himself or Company, to extract peat, **oil** and other minerals in the Camamú and Ilhéos Districts, Bahia Province" (Brasil 1864, emphasis added).

The articles of the mentioned decree have already pointed out the mandatory presence of and mine engineer managing the mineral extraction work,[2] such profession needed, that time, graduated professionals. With the creation of the Ouro Preto School of Mines (1875), Brazil has its Mines Engineering course opened, with the first class formed in 1878.

The articles of the decree also make direct reference to the peat and/or oil, and the extraction cannot start without the instructions and care of a sanitary order, aiming to prevent or remedy any damages caused to the salubrity of the surrounding sites. When treating the used techniques, this decree mentions the machinery, tools, and appliances particularly used in the mining service, and there are no detailed descriptions of the geological conditions of the explored area, but it is highlighted—and it is implied—that the sites described had some

[1]On decrees before 1864, we found terms such as bituminous mineral or words in English such as illuminating vegetable turf.

[2]The term extraction of minerals, in this period, was referring to any type of compound that could be extracted from the soil.

D. Peyerl, *The Oil of Brazil*, Historical Geography and Geosciences,
https://doi.org/10.1007/978-3-030-13884-4_2

potential and they were selected in the attempt of also extracting minerals, as in Camamú, object of the journal *O Auxiliador da Indústria Nacional* (1855) and he mentioned decree (1864).

In 1872, the second decree is published on the exploration of petroleum in Brazil. It is the decree no. 5.014, from July 17, 1872,[3] in which it is possible to observe the use of geographic and geological techniques in the quest for concrete results, mainly regarding the enforcement of presenting geological and topographical plans of the lands explored by the dealers, with geological profiles that show as much as possible, the overlapping of the mineral layers. Although we had no access to documents[4] that proved the compliance of these obligations contained in the decree, what most catches our attention is the geological initiative of detailing the territory, even if such initiative (demonstrated by the articles in the decree), at first, is a form of control of economic interests—interests that were bounded almost directly to the predominant activity of the period: the agriculture.

Other decrees[5] followed the same subject. We raise awareness to the decree No. 8.840 from January 5, 1883,[6] that rises from other technical advances, present in that time, such as the use of drillings, which required previous authorization from the owners of the explored lands.

We highlight that the object of the three presented decrees and other 16 ones that treated the subject, all granted during the Second Reign (1840–1889), which are initiatives of general mine exploration, among which is the petroleum. Although the explorations are concentrated mainly around the mineral oils, peat, and hard coal, which had their consumption in increasingly growth, the fact of discovering petroleum was inserted in the context as a probability, by the association with the minerals abovementioned. Therefore, comprehensively, there is "an uncertain knowledge about the explored land: any mineral eventually discovered would be included in the concession" (Dias and Quaglino 1993: 6). These specific concessions were defined in their fundamental aspects by the 1824 Constitution and by the Land Law from 1850,[7] which established that the "underground was part of the State property and it can be explored by private parties" with mandatory imperial authorization "for the prospection and mining of mineral resources" (Dias and Quaglino 1993: 2).

Thus, we notice that in the Imperial Period, the advances related to the oil quest and exploration were achieved not only by the decrees, but also by transformations that followed new institutions and commissions, which contributed to the development and enhancement of techniques from several scientific areas, such as geology and paleontology, and specifically in the petroleum studies (Peyerl 2010).

Among the commissions created, we mentioned the Brazilian Geological Commission (CGB), operating from 1875 to 1878, and it can be considered "the first institutional initiative, with national reach, in the specific field of geological sciences in Brazil" (Figueirôa 1997: 150). She carried out important studies in the paleontology and paleostratigraphy areas, two important tools to the research of carbonaceous deposits (Figueirôa 1997), which have contributed posteriorly to the oil research.

[3]"It is granted to Luiz Matheus Maylaski a two-year permission to explore hard coal and oil from the Districts of Sorocaba, Itapetininga and Itú, in the Province of São Paulo" (Brasil 1872).

[4]Until the moment, by the research made, the existence of such documents is unknown.

[5]The list of all decrees analyzed in the period (1864–1938) can be found in chronological order in Appendix A.

[6]"It is granted permission to Dr. Gustavo Luiz Guilherme Dodt and Tiberio César de Lemos to explore minerals [including petroleum] in the Province of Maranhão" (Brasil 1883). The decree No. 8.840 does not have the word petroleum in its description; but in its body, the word is key, which is the reason why we use it here.

[7]"The Land Law from 1850, through its Regulation, determined that all lands obtained in sesmaria (property) or through possession—i.e. the lands that were under private domain, should be measured and marked; thus, the sesmarias could be revalidated and the possessions legitimized, assuring the title of definitive property to their owners. The national public lands, named as unsettled lands, could not be obtained simply by occupation, they should be purchased from the Government" (Monteiro 2002: 55).

In 1886, the Geographical and Geological Commission of Sao Paulo (CGG)[8] was created, operating between 1886 and 1931, based on a "line that could be classified as 'naturalist', with the activities covering the fields of Geology, Botanic, Geography, Topography, Meteorology, Zoology and Archeology" (Figueirôa 1997: 113). The contribution from the CGG regarding the petroleum research lasted until the twentieth century, which will be approached later in this text.

Another institution that took an essential part in the formation of professionals, connected to the mining studies, was the School of Mines, founded in the end of 1875 and located in Ouro Preto, Minas Gerais State. The organizer and first director of the School was the French Claude-Henri Gorceix (1842–1919),[9,10] who also was teacher of Mineralogy, Geology, Physics, and Chemistry areas in which the main subjects of the course were centered. Regarding the installation of the School, it almost did not happen, as only three students applied at first, they were:

> […] they sent to the Empire Ministry, José Bento da Cunha and Figueiredo, a petition asking concession of pecuniary assistance in order to continue in the Province of Minas Gerais, for being poor, and without it, they would be forced to give up from their registration. Gorceix comments such request by saying: "The budget for the School in 1876–7 is set in a way that even by adding those expenses, it would still be possible to save some money. But, if by higher order we would not be allowed to help those students and consequently, the opening of the school become impossible, I would be very happy and glorious to be called to enable the setup of an establishment that I had the honor to organize. I ask you, Mr. Minister to withdraw from my wage half of the amount necessary to supply the maintenance of three students in the School of Mines of Ouro Preto, requesting from your generosity the rest of this aid" (Gorceix 2011: 264).

In 1878, the School of Mines graduated the first class (UFOP, s.d.) formed by three mines engineers, integrated by Antônio Veríssimo de Matos Junior, Leandro Dupré, and Francisco de Paula Oliveira,[11] who stands out, as he produced over his life more than 40 works about economics geology of the lead, mercury coal, etc., in addition of studies about the gold deposits in Minas Gerais State, copper in Bahia State, and carbonaceous formations in the south of the country (Ensaios Cronológicos, s.d.).

In the meanwhile, the industrial development in Brazil, relying on the exportation agriculture, was occurring in parallel by enhancements and changes, such as railways, which needed alternatives for the high costs of fuels. Also, the energetic transformations are highlighted, which occurred gradually in other countries, such as the replacement of the coal by the oil, the creation of the diesel engine, electrical illumination, the production of chemical substances, etc. These changes, in big proportions, could be seen and appreciated mainly in the Universal Exposition, which have been happening since 1852[12] in different countries, we highlight the Universal Exposition in March and October 1889 in Paris, in which Brazil took part exposing several objects and awarded in the category of scientific instruments with the piece of equipment Alt-Azimute.[13]

[8]For more information regarding the Brazilian Geological Commission (CGB) and the Geographical and Geological Commission of São Paulo (CGG), read Figueirôa (1997).

[9]He was born in France and graduated in Physics Sciences and Mathematics. In 1874, invited by the Emperor, founded the School of Mines of Ouro Preto. In 1891, he returned to Europe and in 1896 he returned to Brazil to organize the agriculture teaching in Minas Gerais. He passed away in France on September 6, 1919.

[10]The footnotes that describe the biography of the people mentioned, most of times, are formed by many and different general references. For that reason, it was not possible to add the exact reference.

[11]Francisco de Paula Oliveira "started as a mineralogy preparer and repeater of chemistry and physics […]. He managed a few private mining sites, until being used by Derby in the Geological Commission of Sao Paulo, from there he went to the National Museum, then the Commission of Construction of Belo Horizonte, to the public service of the State of Rio de Janeiro, the Commission White, the Commission of the Brazilian Central Plateau. He finally, finished his career as a geologist of the Geological and Mineralogical service of Brazil" (de Carvalho 2010: 97).

[12]A few Universal Expositions during the Second Reign (1840–1889) were made in London (1851 and 1862), Paris (1855, 1867 and 1889), Vienna (1873), Philadelphia (1876), Chicago (1839), among others.

[13]Instrument to determine the height and the azimuth of a world.

These expositions were "an important vehicle of release of the progress achieved by the science, technology and culture" (Freitas Filho 1991: 73), "presenting to the public the stages of the industries, exposing the enhancement of the machinery, also new creations" (Heizer 2009: 13). In this phase of the process of industrialization, the petroleum is presented and seen as a possible and newest source of world energetic source that would result in several discussions about its origin (organic, inorganic, mineral, or even chemical), about its properties and derivative, and about its use as future energy source (Fig. 2.1).

It was also in the Universal Exposition of Paris, in 1889, that the petroleum was greatly highlighted in the geology section:

> Great part of the pages of each edition is devoted to the use and future success of the employment and exploitation of petroleum. The exhibition would contain products at the same time useful to commercial and industrial societies as it would present figures and plants with the aim of highlighting the importance of oil in different parts of the world. For example, the article "in America, a Standard Oil Company has 6,000 km of pipeline"[14] gives a description that emphasizes the history of petroleum, using varied graphical features such as panoramic views, description of uses and applications in industry (Heizer 2009: 12).

The Universal Expositions contributed to the development and expansion of new energetic sources, mainly the petroleum. Also, at that time, there was a development of the internal combustion engines, which provided a growth in the oil industry worldwide. Everywhere where the petroleum was plentiful, the old steam engines were replaced by diesel engines.

As a consequence of the First World War (1914–1918), there was another transformation of this magnitude, of global interest and more intensely, translated in the advance of the industrialization processes that needed "technological research to solve the related technical problems" (Vargas 1994: 214), as the use of petroleum and its derivatives as the main energy source, the exploitation/prospection of petroleum, and the development of the geophysics, which contributed a lot in the discovery of the petroleum. This global scenario had a clear reflection in the initiatives carried in Brazil, which can be seen in detail in the legislation history of this theme that regulated the concession, exploitation, and economical dynamics of the petroleum sector in the country.

That way, both initiatives from natural person and legal person (foreign and national), mentioned in the following items, were connected to the statutory exploratory activities through 74 decrees,[15] executed in the period from 1864, when the word *petroleum* is used for the first time in the description of the decree No. 3.352-A, and it ends on December 17, 1938, when the National Petroleum Council was created. Its end on 1938 is a milestone in political and economic terms, for the research of petroleum by the modifications that the concerned agency had introduced in manpower improvement and specialization sectors, which will be discussed in the third part of this book.

Among the 74°, we observed periods with higher concentration within few years of political changes or when strong government attitudes are taken to find petroleum, and that way, start its exploitation. We highlight that, although these decrees were signed by the Emperor (Empire period) or by the President (Republic period), other acts of concessions were made in the state sphere (Republic period), but in a timider way (reason why we will not talk much about that), as we can demonstrate in the examples below:

> Amzonas State granted, just before the National Revolution from 1930, all the sedimentary area of the Amazon valley in its territory (over 1 million km^2) to 3 foreign companies, for research and exploitation of the underground (Távora 1955: 14).

It can be said that the insistence of private, state, and federal initiatives were essential for the formation of the oil exploitation policy in Brazil. Starting from the 74 analyzed decrees, the

[14]Vargny 1889 apud Heizer (2009: 372).

[15]The selection of these decrees is based in the onset of the word petroleum in their description—with two exceptions, when the occurrence is in the body of the text (decree No. 8.840, from January 5, 1883 and decree No. 393, from May 12, 1890)—as well as the mention of foreign countries that aimed to explore the oil.

Fig. 2.1 The petroleum panorama in the universal exposition from 1889 in Paris. *Source* Barbuy (1999): 114

following graph provides a better view of the years that gather more publications (Graph 2.1).

Following this same line, for a better understanding, we present two maps that complement the graph above. Map 2.1 was elaborated according to the exploitation points mentioned in most of decrees analyzed here.[16] It is necessary to emphasize that the presented points are an approximation of the sites, because, during part of the studied period, the territory demarcations were known as Provinces.

Map 2.2, elaborated with data presented in the work Diniz Gonsalvez, from 1963, with title Petroleum in Brazil (*O Petróleo no Brasil*) (Gonsalves 1963), where we have found a table named "oil drillings made by the Federal Government all over the national territory from 1919 to 1928". Using this table (which can be found in Appendix B of this book), we elaborated a map of the points explored by the Empire/Federal Government, in order to provide a more clear and direct view of the sites and their geographic distribution. Gonsalvez also clarifies, regarding the type of drill used, named *Ingersol Hand*, rotational, which works with granulated steel. "The maximum range, checked for those types of drills, was no more than 600–500 m deep" (Gonsalves 1963: 143).

Both maps showed the exploration points and possible sites where one could find oil, based on the local geological characteristics. It is observed within the next years that the incomplete knowledge about the geology of the territory prevented and frustrated new findings.

2.2 National Legal Measurements Related to the Petroleum (1891–1938)

The Constitution from 1891 "replaces the *dominial* regimen of property of the mines by the accession, assigning the property of the underground and its wealth to the owner of the respective land, as an accessory property" (Távora 1955: 16). In a general context, two principles are highlighted in the legal reordering determined by the Constitution: the establishment that "the land ownership included the ownership of the subsoil" and "it was transferred to the landowner an immense patrimony and for the states the responsibility for the government

[16]It is emphasized that only the name of Monte Negro, in the State of Bahia, is not included in the map, as no references were found to the name of the municipality until the current time.

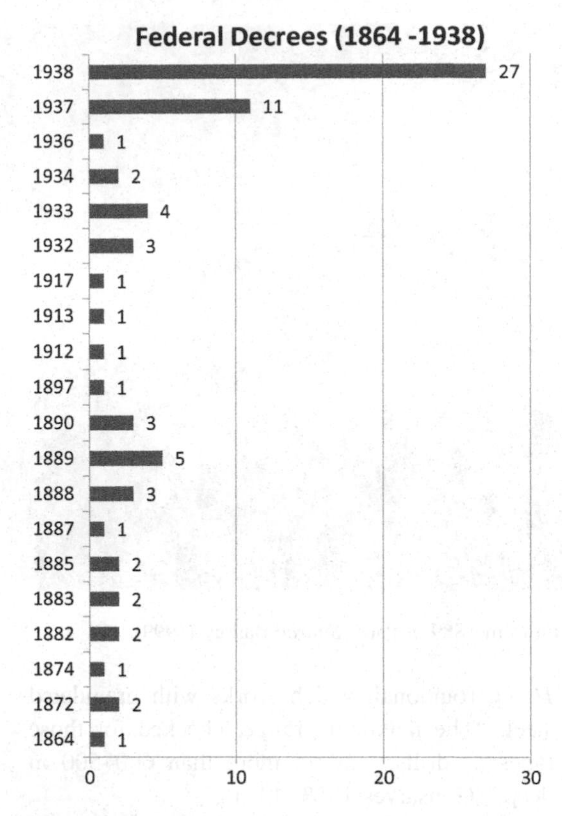

Graph 2.1 Number of Brazilian federal decrees regarding the petroleum (1864–1938). *Source* Elaborated by the author

policy for stimulating mining" (Dias and Quaglino 1993: 7). In addition, states were responsible for "a large part of the responsibility for the granting of exploration licenses and the conduct of geological surveys" (Morais 2013: 39).

In 1903, João Pandiá Calógeras (1870–1934)[17] publishes the Mines of Brazil and their legislation (*As minas do Brasil e sua legislação*), in this work the author defends use of energy resources such as coal and oil and the self-sufficiency of the country related to them. Later, in 1915, the Calógeras Law (decree No. 2.933, from January 06), which was based in his work and was approved when the author was federal deputy, was an initiative that although it was never executed, it "tried to soften the rigid principle of accession, arranging the cases and conditions, which the mineral resources could be exploited by parties other than the surface owner" (Leoncy 1997: 9).

Calógeras was an expert in steel industry; in this period, two fields of the industry were widely growing: the steel industry based on coal and wood and the iron. During the 20s, "the consumption of iron in Brazil almost doubled, due to the use of steel bars[18] in the construction of the reinforced concrete" (Vargas 1994: 25). In this same period, the researches about petroleum were under the Government command of initiatives, which were also involved in the exploration of other natural resources, including the ones bonded to the energetic matters (Fig. 2.2).

Only as an additional information, we directed the look to the situation in the worldwide context, which is very important for the Brazilian economy. In the beginning of the 20s, Brazil was globally seen as a territory with no oil, which had led many foreign companies to concentrate their efforts in making Brazil an importer of petroleum, changing the focus of their investments.

Also, on the same year as the Calógeras Law (1915),[19] the Geological and Mineralogical Survey of Brazil (SGMB), which will be further discussed with more details, is restructured (by the decree No. 11.488, from January 11, 1915), and according to Calógeras:

> [...] it obeyed the technical prescriptions of the soil and subsoil research, it given freedom of movement to the people in charge, it allowed the expansion of the work according to the economic

[17]He was born in Rio de Janeiro. Graduated in 1890, mine engineers with civil benefits by the School of Mines. He was a federal deputy for several candidacies and Ministry of Agriculture (1914) in the Government of Venceslau Brás (1914–1918). One of his initiatives was to regulate the property of the mines and to remodel the pastoral and geological sectors.

[18]Iron with high purity degree.

[19]The Law from 1915 makes no direct mentions of the petroleum, including in the legal definition of mines only the item "mineral oils". Nevertheless, its first title—of the Mines in General—presents a great legal effort to circumvent the absolute rights of the owner of the soil, established in the Constitution. The division of the mine between heirs of the land property was prohibited; the direct administration of the mines by the federal government was permitted if it were the will of the owner and the figure of the "inventor of mines was created" (Dias and Quaglino 1993: 10).

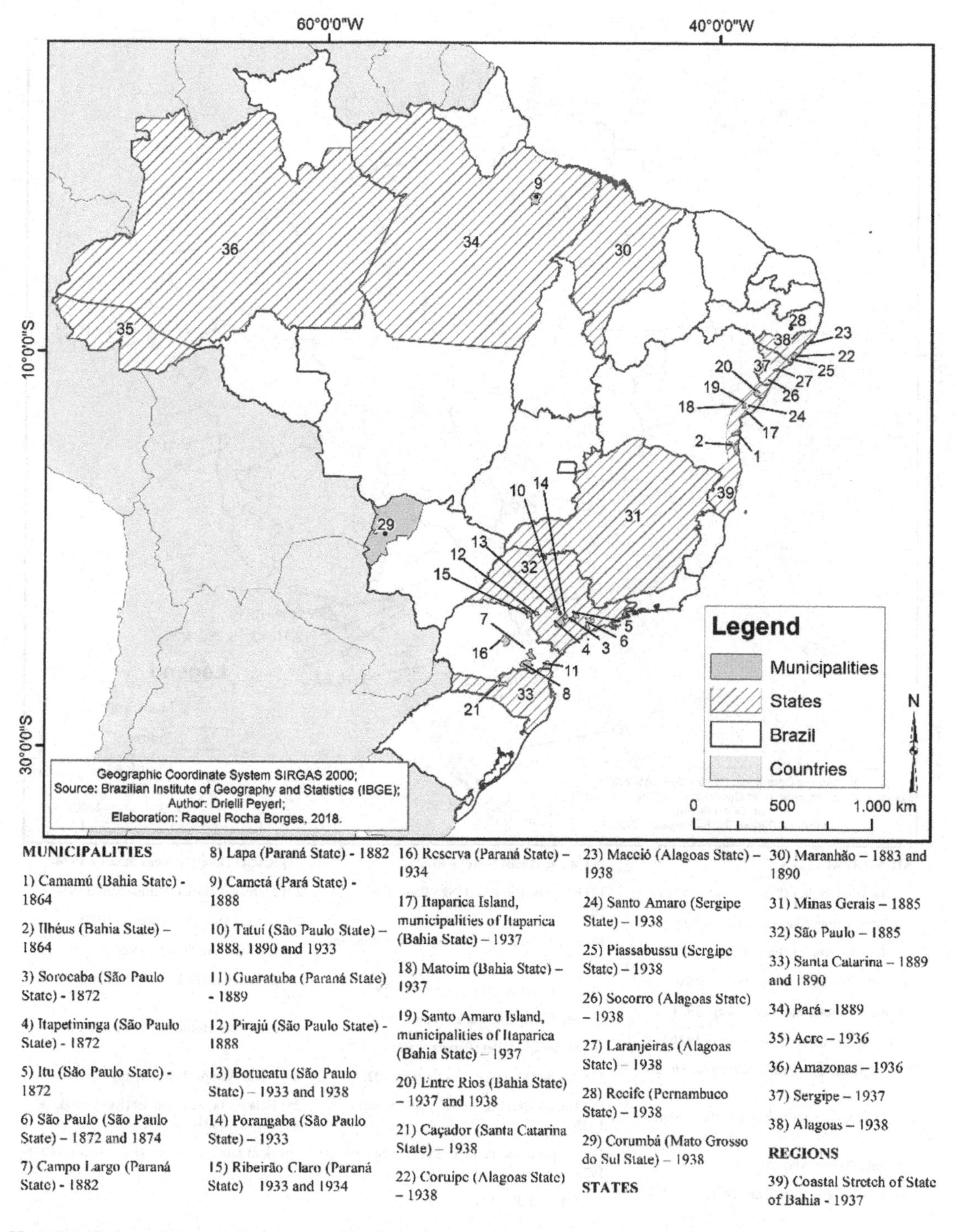

Map 2.1 Points of the petroleum concessions/researches by Brazilian federal decrees (1864–1938). *Source* Borges and Raquel Rocha (2018)

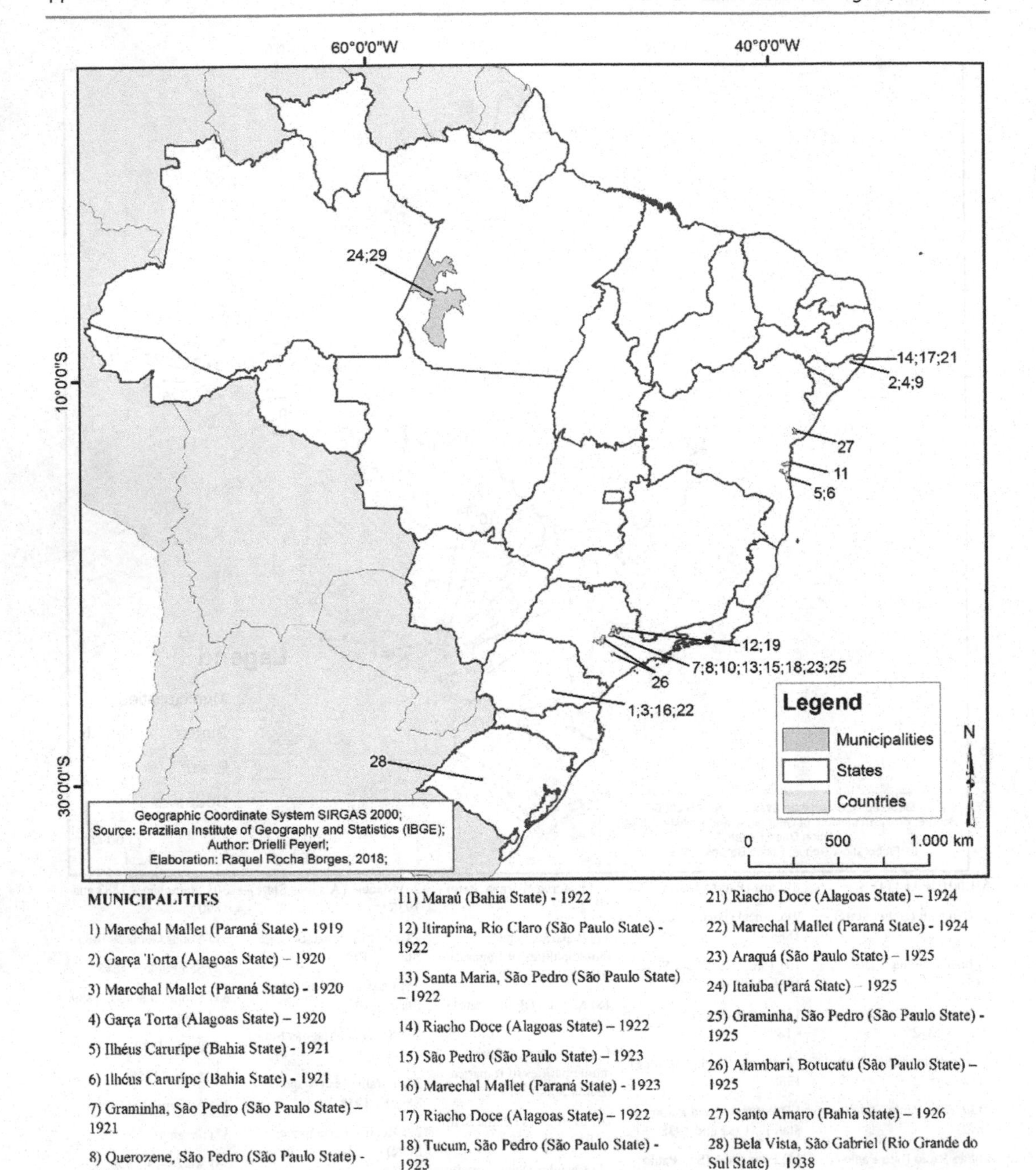

Map 2.2 Petroleum drillings made by the Federal government all over the national territory from 1919 to 1928. *Source* Borges and Raquel Rocha (2018)

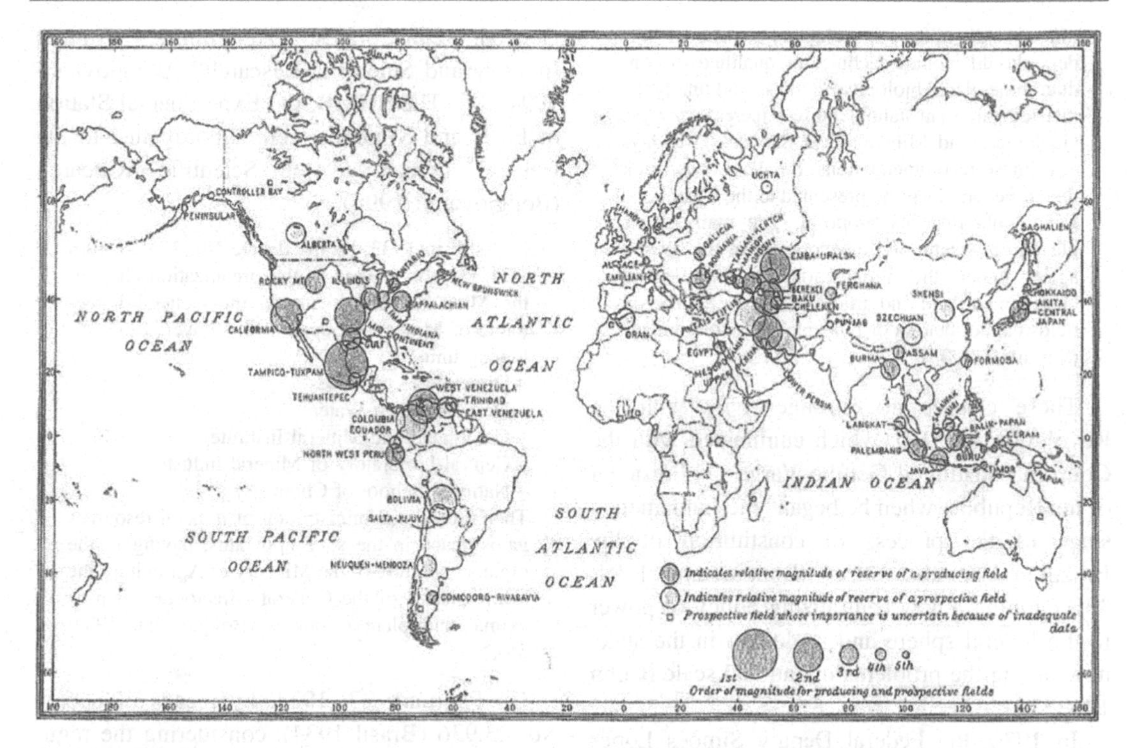

Fig. 2.2 Map of the distribution and proportion of the world oil resources elaborated by the US Geological Survey (1920). *Source* Thom (1929): 135

> needs of the country and the possibilities of the Treasury (Cohn 1968, p. 11).

Ildefonso Simões Lopes (1866–1943),[20] Minister of Agriculture (1919–1922) back then, "effectively supported the activities of the SGMB, and in 1920, 23 engineers were hired" (Figueirôa 1997: 227). In 1921, the Law Simões Lopes widened the power of the State over the mineral resources.

> From the legislation point of view, since the beginning of the twenties, a character is in the spotlight: Ildefonso Simões Lopes.[21] As the Minister of Agriculture, Industry and Commerce, the supported the first initiatives of the SGMB in the exploration of petroleum, promoted the creation of the Fuel Station (decree No. 15.209, from December 28, 1921), bonded to the SGMB itself, and presided the Congress of Fuels, held in 1922. In the area of reform of mining legislation; however, his work was even more prominent (Dias and Quaglino 1993: 14).

Despite the aforementioned initiatives, it was only in 1926 that the Chamber of Deputies presented the first initiative to define an oil policy when the indications of its existence in the country were still vague, rare, and controversial. Such initiative, according to Simões Lopes, would be made to legally arm the country to stop any advance by international companies (Martins 1976).

In June 1927, Ildefonso Simões Lopes, by request of the Ministry of Agriculture, studied the oil issue, recommending the following guidelines to the Federal Government:

> He recommended that the Federal Government increased its activity in relation to oil and the issue of a special law about the petroleum. He asked for the training of petroleum geologists at graduation degree and the overseas training of all petroleum technicians. He believed that the Federal Government should supervise all research contracts and investigate those that were contrary to the national

[20]Brazilian Politician. He was State and Federal Deputy and Minister of Agriculture, Industry and Commerce. He participated in the Commissions of Transportation, Agriculture, Treasury, Special, and Budget.

[21]Brazilian Politician, born in Pelotas City, Rio Grande do Sul State.

> interest. All military installations at national borders should include technicians qualified to conduct mineral and biological studies. And finally, he insisted on a substantial budget increase for the Geological and Mineralogical Service. Two days after these recommendations, Simões Lopes, back then a Federal Deputy, presented to the Chamber a Law containing his proposal. The main clause placed the wealth and development of the subsoil again under the jurisdiction of the Federal Government, and 2nd article said in part: "The oil fields cannot belong or be explored by foreigners" (Smith 1978, 33).

These clauses got stronger only with the Revolution of 1930, which culminated with the Coup that instituted Getúlio Vargas as President of the Republic, when he began "the fundamental stage of the process of constitution of the Brazilian Capitalist State" (Bongiovanni 1994: 33), defining a new centralizing policy of power in the federal sphere and no longer in the state, now taking the problems in national scale (Cohn 1968: 13).

In 1930, the Federal Deputy Simões Lopes defended the previously mentioned measures as immediate and concrete to solve the problem of oil and soil prospecting. Lopes proposed the return of the separation of land and subsoil property, sending Brazilian technicians abroad to specialize them, as well as a reanalysis of contracts between the states and private companies to operate underground and the increase of the budget of the Geological and Mineral Survey of Brazil, making them to devote two-thirds of their resources to the petroleum research (Martins 1976). However, the oil-related controversy spread to the Chamber of Deputies, with several changes and replacements.

The Revolution of 1930 brought changes in the federal administrative plan through a series of measures to streamline the heavy state machine. One of the first measures was the dismemberment of "the Ministry of Agriculture, Industry and Commerce in the Ministries of Labor, Industry and Code and Ministry of Agriculture, to which the agencies connected to the mineral resources were subordinated" (Bongiovanni 1994: 33). In 1933, "the Ministry of Agriculture was structured in a Secretariat of State and three General Directorates: Agriculture, Animal Industry and Scientific Research" (Bongiovanni 1994: 34). The SGMB, the Experimental Station of Fuels and Minerals were subordinated to the General Directorate of Scientific Research (Bongiovanni 1994).

> [...] still in 1933, by the decree No. 23.016 from 7/28/33, it is created in the organization chart of the Ministry of Agriculture, the General Directorate of Mineral Production that covered the following units:
> - Directorates of Mines
> - Directorates of Water
> - Geological and Mineral Institute
> - Central Laboratory of Mineral Industry
> - National School of Chemistry.
>
> Therefore, the administration of mineral resources gains space in the state apparatus, having in the organic structure of the Ministry of Agriculture the same 'status' of the General Directorate, such as animal and plant resources (Bongiovanni 1994: 35).

On February 27, 1934, by means of decree No. 23.936 (Brasil 1934), considering the regulation of the current authorizing regimen for research and development of mineral deposits, as well as the simultaneous authorizations for one and the other activities, one started to separate the authorization for research from the concession for mining, which "implied an increase of state administrative control over mining activity" (Martins, L. apud Bongiovanni 1994: 35). On July 10, 1934, the Mining Code is decreed (decree No. 24.642), considering, mainly, the need to remove obstacles related to subsoil richness, that is, to make the subsoil property independent of that one of the soil:

> As a result, the mineral wealth of a specific area of land was no longer the property of the owner of that area, but they became property of the public domain. By this Code, all the wealth of the subsoil has come to be considered as patrimony of the Union, requiring, to be explored, a special concession from the Federal Government, both for research and mining (Cohn 1968: 17).

For Juarez Távora, the Mining Code:

> [...] facilitated the private initiative for mineral exploration, releasing it on the one hand from the demands of landowners and condominium issues, and, on the other hand, by providing facilities for

> the establishment of soil and subsoil easements, necessary to the exploration, ensuring minimum transport charges and limited taxation, not excessive, together with the financial possibilities of each undertaking. Administrative and technical requirements were created alongside this – these were dictated by the need to rationalize industry and to prevent abuses or omissions that had been demoralizing it (Távora 1955: 28–29).

Thus, prior to the promulgation of the Mining Code in 1934, it is noted that "several foreign geologists, some of whom are surely attached to the Standard of New Jersey, carried studies in Brazil," unhappily those were unpublished works, "except the one of Baker[22] on the basaltic spills of the Paraná Basin" (Abreu 1948: 140).

In addition to the concerns arising from the application of the Mining Code, some works as Bases for the Petroleum Inquiry on Petroleum (*Bases para o Inquérito do Petróleo sobre o petróleo*—1936), Odilon Braga, The Petroleum Scandal (*O Escândalo do Petróleo*), Monteiro Lobato (1936), and The Drama of The discovery of petroleum in Brazil (*O Drama da Descoberta do petróleo no Brasil*—1958) by Edson de Carvalho highlight the constant problems, such as "the lack of resources and the lack of qualified Brazilian technicians as factors that hinder progress in the sector" (Cohn 1968: 22).

In March 1936, facing the problems, the President of the Republic, Getúlio Vargas, assigned a Commission that carried out a rigorous investigation on the official and private action developed in Brazil on the petroleum problem. The Commission was constituted by: José Pires do Rio (1880–1950),[23] former Minister of Transport; Odorico Rodrigues De Albuquerque (M. 1948),[24] Professor of geology at the Escola de Minas de School of Mines of Ouro Preto; Ruy de Lima e Silva (1896–1979),[25] Director of the Polytechnic School; Joviano Pacheco,[26] Director of the Geological and Mineralogical Service of the State of São Paulo; deputy Pedro Demóstenes Roche (1879–1959),[27] mining engineer; General Jose Meira Vasconcellos (1878–1959), indicated by the Minister of War; and commander Ari Parreiras (1890–1945),[28] appointed by the Ministry of the Navy.

One of the first problems investigated by the Commission was the divergence between the leaders of the former Geological Service and those of the current National Department of Mineral Production (DNPM), which would have more results from:

> […] the evolution of petroleum geology knowledge itself, in recent times, then from any incompetence or professional deafness. The greatest of these divergences is manifested among those who admire the possibility of oil in the south of Brazil and those who observe that one has never found oil in the Gondwana rocks of the southern hemisphere (Pires do Rio 1937: 1).

This divergence was manifested in technical reports by foreign geologists (for example, the Commission for Studies on Coal Mines, described later), lasting for decades. The Commission structured by Vargas still suggested a change in the organization of the activity in the field, in which it would distinguish the prospecting of industrial end use prospecting from the scientific study of geology (Pires do Rio 1937). For the Commission, the Mining Code sought to defend the wealth of the subsoil with more emphasis on the local interests rather than protecting them from foreign interests. Then, in 1940 this approach was modified, so that the efforts began to focus on the defense of national wealth against its exploration by foreigners (Pires do Rio 1937), since, in 1939 occurred the discovery of the first oil well in the State of Bahia. Thus, the main requirement of the technicians of the former SGMB was to invest in a "greater volume of resources for the acquisition of drills", in "perforations" and in "preparation of technicians"

[22]Charles Laurence Baker (1887–1979) was a North American geologist.

[23]Brazilian Engineer and Politician.

[24]Graduated in Mining and Civil Engineering. He carried out important studies related to the geology in Amazon and research of coal outcrops.

[25]Civil Engineer by the Polytechnic School of Rio de Janeiro, graduated in 1918.

[26]No biographic references were found.

[27]Graduated in Mining and Civil Engineering.

[28]Brazilian Military and Politician.

(Dias and Quaglino 1993: 16)—which would be difficult to achieve due to high costs.

These uncertainties, contradictions, and resources, however, had been eliminated beforehand with the nationalist laws of 1938 and with the creation of the National Petroleum Council (CNP), all under the new political conditions determined by the New State (*Estado Novo*) in 1937 (Dias and Quaglino 1993). During this period, Getúlio Vargas's policy alleged that "iron, coal and oil would be the mainstays of the economic emancipation of any country" (Vargas 1964: 55). The Revolution of 1930, therefore, changed the Brazilian oil policy, since it began to acquire a certain degree of clarity and gain more obvious nationalist stances through decrees and laws. It is at this point that nationalism assumes a stronger economic, political, and social position. During the New State (1937–1945), one can clearly observe features geared toward nationalism in the political-literary sphere:

> The New State combines a proposal to modernize the restoration project aiming to build a collective identity. The revival of tradition and the search for the true roots found in the regional theme the starting point capable of guaranteeing the integration of the whole national (Oliveira 1990: 195).

However, we are not intended to discuss the use of the terms *nation, nationalist, or nationalism*, often treated as data (in the symbolic level) and not as a problem of ideological dimension within the political, economic, and social dynamics of the country (Oliveira 1990). This is due to the fact that this is not the focus of this book, and it is now only necessary to note that this subject is the object of a vast bibliographical production of magnitude and deepening, such as the one produced by the sociologist Lúcia Lippi Oliveira, The National Matter in the First Republic (*A Questão Nacional na Primeira República*) (1990), by showing the national matter and its definitions in the so-called First Republic (*Primeira República*), or even by Gabriel Cohn, in the piece of work Petroleum and the Nationalism (*Petróleo e o Nacionalismo*) (1968). I complement this discussion with the observation made by Eric Hobsbawm: "The State does not limit to define the coordinates of the national problem: it is the one that many times creates the nation itself" (Hobsbawn 1985: 25).

2.3 Technical and Empirical Development in the Search for Oil (1897–1939): The First Deep Drillings

The first deep drilling for oil that has been reported was in 1897, sponsored by the farmer Eugênio Ferreira de Camargo (1869–1919),[29] who invested in the exploration and exploration of oil in Bofete Municipality (São Paulo State), and to do so he hired the Belgian naturalist Auguste Collon (1869–1924).[30] They both started "drilling in the Morro do Bofete, in Tatuí City (São Paulo State), 150 km from the capital" (Oliveira 1940: 18), which was concluded in 1901, and achieved 448.5 m, and only sulfurous water. This research results in a manuscript full of details from Auguste Collon, with the title "Le Petrole dans les Environs du Mont de Bofete et de Porto Marints", considering as initial landmark of the technical and scientific works about petroleum in Brazil (Felicíssimo Júnior 1970 apud Oliveira 2010).

Another known drilling was made almost a decade after that, in 1906, around Guareí City (São Paulo State), made by the Geographical and Geological Commission of Sao Paulo, achieving 139 m deep (Mentira Velha ..., 1961: 11). It is noteworthy that most of the researches conducted in the late nineteenth and early twentieth centuries were destined to find coal, one of the main energy sources at that time. Specifically, in this book, we will only approach drillings directly related to petroleum.

[29] Born in Campinas, he was a reach farmer, known as pioneer in the exploration activity in the country. Together with the Belgian naturalist Auguste Collon, they were responsible for the first deep drilling.

[30] Auguste Collon was born on April 30, 1869, in Mons, Belgium, and studied in the Athenée d'Ypres, where he graduated in the Humanities course. He also attended from 1885 to 1891 the Ph.D. in Natural Sciences in the University of Liège. At the age of 27 years old, he arrived at Porangaba, hired by the farmer Eduardo Ferreira Camargo, returning to his homeland in 1897.

In 1917, it was created the Commission of Drilling and Research of Stone Coal and Petroleum in the Amazonas Valley, headed by the engineer Antônio Rodrigues Vieira Junior.[31] This Commission resulted from the studies made by the Geological and Mineralogical Survey of Brazil (SGMB), created in 1907 and headed by Luiz Felipe Gonzaga de Campos,[32] from 1915, after the suicide if Orville Adelbert Derby (1851–1915).[33] Gonzaga de Campos carried meticulous studies in the Amazon Basin in 1913, 1914, and 1915, which resulted in the creation of the referred Commission (Mentira Velha …, 1961: 11):

> The work of this Committee continued in 1918 under the direction of the engineer Avelino Inácio de Oliveira[34]; in 1919, headed by the engineer Paulino Franco de Carvalho[35]; in 1922, headed by the engineer Egeo Marino de Almeida[36]; in 1924 under the direction of the engineer Pedro Moura, in 1928 headed by the engineer José Lino de Melo Júnior[37] and finally in 1929 under the command of the engineer Antonio Moreira de Mendonça[38] (Mentira Velha …, 1961: 11).

From 1917, there are works on deep drilling directly related to petroleum. Another example, besides the aforementioned, would be occurring in the state of Sao Paulo through a private company, which received drills from the government. This would be the Petroleum Company of São Paulo State, which carried out drilling at Rio Claro (São Paulo State) between 1917 and 1918, reaching 300 meters deep. According to Glycon de Paiva[39] "the drill belonged to the government and the drilling was followed by its technicians" (Paiva, G. apud Oliveira 1940: VI).

In 1918, the company requested, in addition to the material and technical assistance offered by the government, the defrayal of the machines, but the request was refused because of the high expenses. The company then abandoned the work and it was up to SGMB to assume both the costs and the drilling. The first federal drilling, that is, the one carried out under full responsibility of the government, with the intention of finding oil, was carried out in Marechal Mallet (Paraná State) in 1919 (Paiva, G. apud Oliveira 1940). In this period, it was also considered to contract foreign companies to carry out drillings together with those already carried out by the SGMB due to the difficulty in finding oil (Dias and Quaglino 1993).

The President Epitácio Pessoa (1919–1922),[40] in a message to the Congress pointed out that the cause of the failure of the oil-seeking investigations hitherto made was tied to incomplete knowledge of the geological structure of the country, and proposals and requests from foreign companies had recently appeared for the exploration of oilfields (Cohn 1968). However, even with the proposals and requests of foreign

[31]No information was found about his biography.

[32]Graduated by the School of Mines of Ouro Preto, in the class of 1879, mining and civil engineer. He worked with Derby in the CGG and SGMB. He studies the carboniferous region of Santa Catarina State. He carried out a survey of iron reserves in the central region of Minas Gerais State. He tried to use the schist from Maranhão State in the production of gas in Sao Luis. He was the creator of the Experimental Fuel and Mineral Station, which gave rise to the National Institute of Technology (1933).

[33]North American geologist naturalized Brazilian. He was born in Kellogsville, New York State and passed away in Rio de Janeiro. He joined the Brazilian Geological Commission (1875) and directed and founded the Geographical and Geological Commission of Sao Paulo (1886) and the Geological and Mineralogical Service of Brazil (1907).

[34]Graduated by the School of Mines of Ouro Preto, in the class of 1916, mining and civil engineer. He worked in the SGMB and was the director of the Mineral Production Promotion Service from the National Department of Mineral Production. He published several papers regarding the research done when he was in the Geological Service. Among them, the book Geology of Brazil (Geologia do Brasil), published in collaboration with Othon H. Leonardos (1899–1977), Brazilian engineer and geologist.

[35]Graduated mines and civil engineer by the School of Mines of Ouro Preto, in the class of 1918.

[36]Full Name: Egeu Marino de Almeida Gomes. Graduate mines and civil engineer by the School of Mines of Ouro Preto in the class of 1921.

[37]Graduated in 1924 in Mines and Civil Engineering by the School of Mines of Ouro Preto.

[38]No information was found about his biography.

[39]Glycon de Paiva Teixeira graduated in 1925 in Mining and Civil Engineering by the School of Mines of Ouro Preto.

[40]Epitácio Lindolfo da Silva Pessoa (1865–1942) was a Brazilian politician.

companies, the government of Epitácio Pessoa did not adopt the idea, making it clear that "minerals in general, and oil in particular, would be exploited exclusively by Brazilians or would not be exploited at all" (Smith 1978: 27).

In 1921, the geologist Euzébio Paulo de Oliveira, in charge of the SGMB, was in charge of all the oil drilling carried out in the country. Before attempting to solve the oil problem (drills), Oliveira tried to develop his own techniques instead of hiring them, trying to form petroleum geologists and Brazilian drillers (Marinho júnior 1989: 217). Previously, he had already set goals for petroleum research in Brazil, as read in Bulletin No. 1, from 1919: "(1) selection of drilling sites by national geologists; and (2) perforations by Brazilians" (Smith 1978: 28). We emphasize that the Oliveira attitude was positive for the entry of Brazilians interested in this area of drilling and petroleum in the SGMB. It was a pioneering initiative of empirical training and practice of petroleum engineers/geologists in the country through the support of people linked to a federal body. Regarding the formation of drillers, Oliveira found greater difficulties: the drills were imported, generally from the United States, having little information about their operation; therefore, their operation was carried out in an empirical way, by means of the accompanying manuals or with the help of some foreigners who were part of the staff of the SGMB.

The Brazilians not only got into contact with foreign knowledge and techniques, but also used their help and manuals. An example of this is that of Mark Cyril Malamphy (1902–1957), a geophysicist who, since December 1931, has already served the SGMB. On February 21, 1932 (Brasil 1932), the decree No. 21.079 authorized the federal government to hire a geophysicist—specifically, Malamphy—for the SGMB:

> The Head of the Provisional Government of the Republic of the United States of Brazil, taking into account the need to develop the research work on the subsoil and on the feasibility of applying geophysical methods in the research of structures suitable for the accumulation of oil and exploration of metalliferous deposits in the country.
>
> Given that these investigations can only be entrusted to technicians who have already demonstrated special knowledge on the subject, and that is why the American Expert Mark Cyril Malamphy, who has been collaborating with the Geological and Mineralogical Brazilian Service since December of 1931 [...] (Brasil 1932, s.n.).

At this point, the direction of the SGMB was still in the hands of Euzébio de Oliveira,[41] who had already acquired an *Oertiling*[42] magnetometer, later used by Luiz Flores Moraes Rêgo (1896–1940)[43] in research in the State of Bahia (Abreu 1948). Malamphy, in addition to start his studies in Geophysics, prepared a group of disciples, such as Henrique Capper Alves de Souza,[44] Irnack Carvalho do Amara (1905–1986),[45] among others (Pinto 1988).

> The geophysical works were inaugurated by Mark Malamphy in 1933, who instructed and trained several engineers of the National Department of Mineral Production and with them performed magnetic and gravimetric work in São Paulo, Parana, Santa Catarina and Alagoas (Abreu 1948: 140).

Thus, in Brazil, two motives were moving in parallel: the first, "ideological (the 'national dignity'), because of the political and economic need to find oil without the presence of foreigners; and the second, virtually technical character, taken at the level of the most immediate 'empiricism' (the subsoil is examined, and the

[41]Director of the SGMB from 1922 to 1925 (Acting Director) and from 1925 to 1933.

[42]The magnetometer determines the strengths of the local magnetic field.

[43]Luiz Flores Moraes Rêgo graduated in Mining and Civil Engineering by the School of Mines in 1917 and joined the SGMB. He contributed to 80 written articles, dealing with paleontology, physical geography, soils, geology, stratigraphy, etc. He devoted himself to fuel problems such as oil and steel problems. He was a Professor of Geology and Minerology at the Polytechnic of São Paulo.

[44]More than 25 published works, with emphasis on Occurrence of Molybdenum in the State of Ceara (Ocorrência do Molibdênio no Estado do Ceará) and The Gold and Life in Some Regions of Brazil (O Ouro e a Vida em Algumas Regiões do Brasil).

[45]He was born in Rio de Janeiro. He graduated in Mining and Civil Engineering by the School of Mines of Ouro Preto in 1931. He held the position of Director of Petrobras twice (1954–1957 and 1961–1963).

results are observed)" (Cohn 1968: 12). Thus, Brazilian professionals from the geology and engineering areas worked together to develop techniques such as drilling (Dias and Quaglino 1993).

In this period, it was necessary to import professionals, drills, and technology for soil drilling, since Brazil was still in the transition from agroindustry to industry development. In addition to the pieces of equipment and the need for specialized professional training, as mentioned, another important factor was the geological constitution of the territory, a factor described in the example that follows about the State of Alagoas.

The state of Alagoas was one of the first, along with Paraná, Bahia, and São Paulo States, to conduct deep drillings. What we want to emphasize is the history of geological studies carried out locally and to point out that the search for oil was based on previous descriptions, combined with the related difficulties and the physical conditions of the territory (transport, location, and geological knowledge available on the area). Research had already been carried out in Alagoas State by Charles Frederick Hartt (1840–1878),[46] in 1886, and by John Casper Branner (1850–1922),[47] in 1901, who described several outcrops of bituminous shales on the coast of that state. Orville Adelbert Derby, in 1907, also referred to these same shales, as well as those of Maraú (Bahia State) (Lange 1961). From 1918, geological studies were carried out in Alagoas State by SGMB technicians. The beginning of the drilling, in 1920, was based on the historical and in previous studies made in the site, not being made randomly.

The first oil well drilled in Alagoas State was located in Garça Torta. Its drilling began on April 14, 1920 and was abandoned on July 9 of the same year, with 78 m deep. Between 1921 and 1927, SGMB drilled, in this state, two additional wells in Garça Torta and three in Riacho Doce. The total drilled in these six wells was 870 m. In all of them, evidence of oil was found, but no commercial production was achieved (Lange 1961).

It is worth pointing out that the commercial value/production of a well depends on a number of local or regional factors. Among these factors, the well geographical location, the presence or absence of facilities for storage in the area, the transport of the oil to the refineries, the cost of drilling to reach this horizon, the recoverable reserve, the size of the field, and the number of wells necessary for its development influenced in the classification of the well, among other factors with economic or technical importance (Lange 1961):

> In 1928, the SGMB established new doctrine on oil prospecting and has replaced the "preferably indexed instruction" with a "scientifically based" approach whereby "all surveys should be carried out where there is a suitable geological structure for the concentration of petroleum in an exploitive quantity". Within the new instruction, drilling continued in 1929 in the States of Pará, São Paulo and Paraná (Marinho Júnior 1989: 223).

In 1932, the National Oil Company began operating in the Riacho Doce area,[48] which began drilling with a drill loaned by the Geological Service to the Government of Alagoas. In 1936, a team from the Mineral Production Promotion Survey (SFPM)[49] proceeded to a geophysical research, delimiting a structure in Ponta Verde, which, however, was never drilled by such service (Lange 1961):

[46] American–Canadian geologist. Coordinator of the Geological Commission of Brazil.

[47] North American geologist. He has published important works on Brazil, such as *Geology and Physical Geography of Brazil, As possibilidades de petróleo e jazidas de diamante [The Possibilities of Oil and Diamond Quarries], Estudo de Paleontologia da Amazônia [The Study of Paleontology in the Amazon] and Os minérios de Manganês [The Northeast, Manganese Ores]*, among others.

[48] We will refer to the National Petroleum Campaign in the item "Brazilian Companies".

[49] In August 1933, with the enlargement of the SGMB, under the new designation—General Directorate of Mineral Production (current National Department of Mineral Production)—the drilling service was transferred to the Mining Department (current Division of Mineral Production Promotion) (Oliveira 1940: 25).

From 1935 to 1937 important works of recognition were developed in Acre under the direction of Pedro de Moura, who advises to return the views now to the studied region. In 1937, by suggestion of Glycon de Paiva, Amaral and Abreu, and with the support of Guilherme Guinle[50] geophysical works were carried out in the Recôncavo, by the engineers of the Mineral Production Promotion headed by Irnack do Amaral. Magnetic profiles were made between Salvador and Feira, on the Ilha de Itaparica, Matarandiva, Valença, and Ilha do Tinharé. Studies with torsion balance and with seismograph were made in the Camaçari region (Abreu 1948: 140).

In 1939, the responsibility for oil research was passed to the National Petroleum Council (CNP), created in 1938. Altogether, 163 drillings were carried out by the SGMB and/or National Department of Mineral Production (DNPM). In 1939, the two drillings conducted by DNPM in district of Lobato, Salvador City, Bahia State, numbers 153 and 163, reached "depths of 71.91 m and 228.38 m", respectively, (Oliveira 1940: 29) that the later found the long-awaited petroleum, "the first that can be designated as an indicator of a petroleum field with commercial value" (Oliveira 1940: 29).

The depth of the first deep drilling presented in the Organizational Chart 2.1 oscillated a lot, depending on the land explored and the available piece of equipment. In 1900, explorations in the United States, through rotary drilling, were already reaching 317 m deep (Mansano 2004).

Unlike Brazil, in the late nineteenth and early twentieth centuries, the United States joined the practice of exploration by means of so-called *wild cat* drillings, randomly, but by the thousands, searching hard for oil throughout the subsoil. Statistically, out of every 100 drillings of this type made, 98 were fruitless (Maya 1938). Drilling by the percussion method went through its golden age in the late nineteenth century, especially in the United States. At the same time, it began to develop the rotary drilling process (Souza and Lima 2002).

In 1900, in Texas, the American Anthony Lucas,[51] using the rotary process, found oil at 354 m deep. This event was considered an important milestone in rotary drilling and oil history. In the following years, the rotary drilling developed and progressively replaced the drilling by the percussion method (Souza and Lima 2002: 8).

In Latin America, in 1907, Peru already had 305 oil-producing wells and, in 1913, this country persisted and put into operation another 102 wells (Souza and Lima 2002). The drillings in Peru, in certain places, reached 1,200 m. However, it is noteworthy that in this period, there was a strong governmental and foreign investment, and the territorial extension, much smaller than in Brazil, made the work easier.

In Brazil, a criticism related to the drilling process was frequent in the country. According to the journalist, lawyer, and Brazilian politician Emilio Maya (1906–1939), the drillings were interrupted when they achieved 100–200 m deep, "and in those areas was necessary to drill 800–1.000 m" (Maya 1938: 93). However, most of the drills[52] used was by the percussion method (*National Supply Co.* and *Keystone*) and rotary (*Ingersoll Rand*, among others), imported from the United States, much of it already used, and could reach approximately 700 m deep, which partially contradicts the words of Maya (1938). No information is available on whether the drills capable of reaching such depth, such as those mentioned, were interrupted by the geological conditions of the territory. It is also important to mention that rotary drilling was the best way to find petroleum, at that time, by reaching great drilling depths and diameters in Brazilian sedimentary basins.[53] In 1939, in Salvador, one has

[50]Graduated in Civil Engineering by the Polytechnic School of Rio de Janeiro in 1905. He got involved in the construction of hydroelectric plants in Bahia. In Rio de Janeiro he founded the Boavista Bank. In the thirties, he made investments in the petroleum exploration. Nationalist, financially supported the national liberation alliance, aimed at combating fascism and imperialism.

[51]Anthony Francis Lucas (1855–1921) was an oil explorer.

[52]See Appendix C.

[53]Oil is found in sedimentary basins, "[…] which are depressions on the surface of the soil filled by sediments that become, in millions of years, sedimentary rocks", as the processes of generation, migration, and accumulation of petroleum operate in geological time scale (Gusmão 2005: 179).

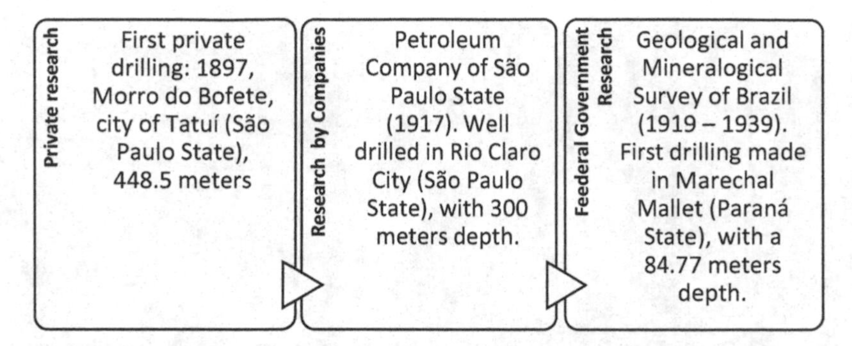

Organizational Chart 2.1 First deep drillings (The first drilling made by the SGMB, in Mallet, occurred in 1919). *Source* Elaborated from Gonsalves (1963, p. 143) and Paiva G. apud Oliveira (1940, p. v)

news of the start of operation of the first rotary oil well drill, known for its fast operation and low costs (Fig. 2.3).

In the case of Alagoas State, on May 29, 1939, the beginning of the drilling in the well AL-1, in Ponta Verde that was concluded on August 20, 1940, with 2.143 m deep. Subsequently, a series of training tests were carried out in 10 areas with oil traces, from which a total 15 barrels were recovered by surging[54] (Lange 1961).

It is worth to remember that the oil exploration/exploration process is time-consuming, slow and expensive, and that three steps are necessary: the geological recognition of the areas to be exploited; geophysical works for the selection of drilling points; execution of drillings to verify the existence of the oil (Peixoto 1957: 275). Another factor to be pointed out was the difficulty, in the mentioned period, of transporting the drills and, often, to access the point for drilling.

Through drillings it was also possible to find another resource: natural gas. The case occurred in Marechal Mallet (Paraná State) in 1922: in the course of a search for coal, natural gas was found (Gonsalves 1963: 24). The drillings at Marechal Mallet continued in the following year, reaching 510 m deep, where a natural gas deposit was found at 505 m deep. According to the geologist Alpheu Diniz Gonsalves (1883–1973),[55] the gas was in a "layer of dark shale, which disintegrates promptly as the water is withdrawn from the borehole and smashes, clogging the hole and preventing the leakage of gas" (Gonsalves 1963: 24). As it can be seen, "the results of oil drilling in 1922, if not positive in the discovery of liquid fuel, have, however, proven the existence of an unknown mineral resource in the country—natural gas" (Gonsalves 1963: 24).

2.4 Private and Government Initiatives of Oil Research in Brazil (1864–1938): A Panorama

During the second reign (1840–1889), we found, in the decrees that were earlier mentioned in this book, 14 concessions to private oil-related initiatives. The Brazilians and/or foreigners who acquired this concession were generally involved in a relevant way in the social and political environment, being colonels, bachelors (lawyers), politicians, or people engaged in the construction of railroads over the country, as we can see below:

- 1864, Thomaz Denuy Sargent (English); 1872, Luiz Matheus Maylaski (noble, Portuguese,

[54]Surging is the well development/stimulation process in which the water is moved by successive vertical movements of a piston immersed in it (Perfuração, s.d.).

[55]Geologist and civil engineer. He was a Professor at the Polytechnic School of Bahia and accompanied the geologist Orville A. Derby in the study of the geology of the State of Bahia. Important character of geology, but we have little information about him.

Fig. 2.3 Use of the first drill rotary oil well in Salvador city (Bahia State—1939). *Source* o petróleo …, s.d

born in Slovakia), known as *Visconde de Sapucaí.* He had been involved in the construction and inauguration of the Company of the Railroad (Estrada de Ferro Sorocabana); 1872/1874, Cyrino Antonio de Lemos (bachelor) son of João Antônio de Lemos, pioneer in the industrialization of hats in Brazil; 1872/1874, João Batista da Silva (politician, lawyer) participated as Lecturer of the Popular Conferences of the Fshreguesia da Glória (1873–1883) in subjects mainly related to agriculture, being one of his participations in 1875 along with Charles Frederick Hartt; 1883, Gustavo Luiz Lemos Dodt (German); 1883, Tiberius Cesar de Lemos (Brazilian and bachelor); 1885, Manoel Vidal Barbosa Lage (politician and farmer from Minas Gerais); 1887, Henri Raffard (major, first secretary of the Brazilian Historical and Geographical Institute, 1895).

After the Proclamation of the Republic (1889), only one grant of private initiative, from 1890 (Brasil 1890), was held. Other preexisting concessions were modified by the addition of new observations, by extension or by the inclusion, together with the name of the grantor, of the words company or society, demonstrating changes in the economic practice of Brazil. From 1897, we have mastered private initiatives through foreign companies, which entered in de decade of 1910.

Only in 1932, after the Revolution of 1930 and in the middle of the Transitional Government (Governo Provisório) of Getúlio Vargas (1930–1934), the concessions are granted again to individuals, but with the replacement of the term "concession" for "authorization" and the term "exploitation" by "research of petroleum".[56] From 1937, the Federal decrees related to oil are exclusively destined for Brazilians, and always begin as follows: "One authorizes the Brazilian citizen […]".

The 30s are marked by the great protectionism and intervention of the Vargas Government,

[56]"If the substantive law, which governs the *Institution* of property […], considers as being private property some mine, the wealth of the soil or the fall of water […], *authorization* is required. If the substantive law, which governs the *institution* of property, or the constitutional text considers *property of the State*, or *common*, the mine, the wealth of the soil, or the fall of water, the *concession* is due" (Miranda 1960, p. 518).

based on the argument that the country could let the foreign companies to take control of energy resources, such as the oil. For that reason, such nationalist attitude is visible in Federal decrees. Although the decrees authorized only Brazilian citizens, the presence of foreign investments continued to be present in the country. The fact is that the nationalism of Vargas government has shown itself to be moderate, due to the high investment possibilities from foreign companies and the attitudes of local nationalist politicians.

Several names of this phase significantly contributed to the country petroleum research and are mentioned in the decrees. For example, Decree No. 1.849, dated from August 3, 1937, authorizes "the Brazilian citizen Silvio Fróes de Abreu[57] to research petroleum and natural gas in an area of 175.84 hectares on the Ilha de Itaparica, municipality of Itaparica, Bahia State" (Brasil 1937). It is emphasized that Silvio Fróis de Abreu (1902–1972) was one of the initiators of the research in the district of Lobato (Bahia State), where petroleum was found for the first time in 1939.

Regarding foreign companies, at first, it is in the second decade of the twentieth century that two decrees formalize the participation of two of them in Brazil, the Standard Oil Company of Brazil, in 1912 and The Anglo Mexican Petroleum Products Company, in 1913. However, the presence of the Standard Oil Company of Brazil predates the decree No. 9.335 from January 17, 1912. The decree No. 2.471, dated of March 8, 1897, grants authorization for the operation of the Industrial Petroleum Company in Brazil, which later came to be called the Standard Oil Company of Brazil in the text of the decree of 1912.

In order to understand the entering process of an American company in national territory, it is necessary to understand its process of creation and transformation. According to Morais, the main event in the world oil industry "in the nineteenth century, in January 1870, when five businessmen, headed by John D. Rockefeller (1839–1937),[58] founded the Standard Oil Company in Cleveland, Ohio" (Morais 2013: 33). In 1879, the company "controlled 90% of the refining capacity of the United States. They also controlled the oil pipelines and collection system of Oil Regions and dominated the transport" (Yergin 2010: 28). They eventually became the largest oil company in the world, forming the petroleum industry's tripod: producing, transporting, and refining it.

Subsequently, the company underwent investigations by the United States Congress and Legislative Assemblies from several states operated by the company, under the suspicion of receiving discounts on railroad projects and practicing methods of restricting the free trade (Tarbell 2003). In 1911, the company has a partial closure of its activities: the Supreme Court of the United States decides to dismantle the monopoly and orders the creation of 34 new smaller companies.

Dependent of the former Standard Oil, the Petroleum Industry Company, mentioned in the 1897 decree, was a corporation founded and organized under the laws of the State of West Virginia. The 1911 decree mentions the name of the President of the Standard Oil of New York, Henry Clay Folder (1857–1930),[59] and the secretary of state of West Virginia, Stuart Felix Reed (1866–1935).[60] The Standard Oil of Brazil renewed its authorization to continue operating in the country by means of six further decrees No. 234, from July 17, 1935; No. 4.894, from November 20, 1939; No. 21.608 from August 20, 1946, No. 30.339, from December 24, 1951. No. 31.472, from September 18, 1952; and

[57]Brazilian chemist and geographer. He was the founder of the National Geography Council of the Brazilian Institute of Geography and Statistics (IBGE). He wrote more than 40 papers, among them: *Xisto Betuminoso da Chapada do Araripe [Oil Shale of the Chapada do Araripe], O Nordeste do Brasil [Northeast of Brazil], Recôncavo da Bahia [Recôncavo of Bahia], and O Petróleo de Lobato [The Petroleum from Lobato].*

[58]John D. Rockefeller (1839–1937) was an American investor and founder of the Standard Oil Company, which was the first major US trust.

[59]He began to work, in 1881, in a position of confidence in Standard Oil. He was the first president of the Standard Oil of New York, serving until 1923.

[60]American politician who represented West Virginia. He was Secretary of State of West Virginia from 1909 to 1917.

No. 31.811, from November 20, 1952.[61] In addition to the operating permits issued by the Standard Oil Company of Brazil, the company also worked in other areas of interest, such as the construction of a gas tank at the Port of Santos, the supply of oil to the Central do Brasil railroad, etc.

Another foreign company to have concessions in the country was The Anglo Mexican Petroleum Products Company, found in 1909. The first concession, in 1913, gave the company permission to operate in Brazil. In 1917, another decree authorized the replacement of the Company's name to Anglo Mexican Petroleum Company Limited, which had also sought, since the first decade of the twentieth century, to settle in other Latin American countries, as in the case of Argentina, specifically in Buenos Aires (Solberg 1979).

The company also worked on the construction of tanks on the Barnabé Island for the kerosene deposit intended for The Caloric Company[62] and Anglo Mexican Petroleum Company Limited, including enclosure walls, platform, pump houses, washing sheds and filling of drums, pipes and belongings (Brasil 1933). It is noteworthy that in this period, in Mexico, both exploration and refining were extremely connected to the United States.

We emphasize, however, the interest of Brazilian companies in finding oil. In 1932, the scenario of the federal decrees related to oil began to change as a result of the authorizations for the formation and organization of Brazilian corporations focused on oil exploration in the country. Decree No. 21.415 from May 17, 1932, opens this period authorizing:

> […] the incorporation by the Mrs. J. B. Monteiro Lobato,[63] M. L. de Oliveira Filho[64] and L. A. Pereira de Queiroz[65] of a joint-stock company, headquartered in Sao Paulo and capital of 3,000: 000 $0,[66] exclusively national, with the objective of researching petroleum formations and exploring their deposits (Brasil 1932, s.n.).

The research should be carried out by prior authorization from the Federal Government on lands with area that does not exceed 4000 hectares. Another decree from the same date (No. 21.414) authorized the same "Brazilian oil Company to continue in the contracts for the assignment and leasing of subsoil of territorial properties in the municipality of Piraju, State of Sao Paulo" (Brasil 1932).

Also known as Company Petróleos do Brasil, the company began to launch actions to raise funds, and the unusual procedure began to arouse the attention of the authorities (Dias and Quaglino 1993). However, at first, Monteiro Lobato and/or Company Petróleos do Brasil interest initially focused on the state of Alagoas and the region of Sao Pedro (SP). According to Monteiro Lobato, the point where there might be oil in greater quantity and a better strategic condition for export was in the state of Alagoas (Sacchetta, s.d.).

Even so, the largest investments were in the wells on Sao Pedro region (São Paulo State), in 1932, when drilling began in the Araquá field[67] (Chiaradia 2008). It should be noted that Petrobras, during its operation, faced difficulties in obtaining permits by means of decrees (state and federal) for the exploration of the territory in search of oil.

In the book, *The Petroleum Scandal* (*O Escândalo do Petróleo*), 1936, Monteiro Lobato compares the State of Sao Paulo with Alagoas: "It is easy to influence fat people, because fat people have fat to lose. The Alagoan is skinny, dry, tightened by the terrible sun of the northeast" (Chiaradia 2008: 195). The author presented the "state of São Paulo as a wealthy state, which did not care about the search for new wealth, while Alagoas, a poor state, sought alternatives that would sustain it—in this case, the search for oil" (Oliveira 2010: 73). The book

[61]In Decree No. 31.811, from November 20, 1952, the change of company name, now Esso standard do Brazil Inc.

[62]American company.

[63]José Bento Monteiro Lobato (1882–1948) graduated in Law and he is considered one of the most influential Brazilian writers of the twentieth century. He played an active part in petroleum scenario, including his search and works on the theme, such as The Oil Scandal.

[64]No biographic information was found.

[65]No biographic information was found.

[66]Three thousand contos de reis (local currency).

[67]Today, city Águas de São Pedro (São Paulo).

also discloses "the relationship of several land concession contracts for research and exploration of Brazilian subsoil wealth to companies organized under national laws, some of which had, behind them, foreign guidance and capital support" (A polêmica ..., 1961: 19).

It is noteworthy that Monteiro Lobato business choice for exploration in Alagoas was linked to a geophysical survey carried out in 1936, by order of the Alagoan Government, made by a German company whose conclusions were favorable to the existence of oil in that area (Carvalho apud Cohn 1968). In addition, Monteiro Lobato was also involved in the formation of other oil companies in Brazil, such as Companhia Brasileira de Petróleo Cruzeiro do Sul (Brazilian Company of Petroleum Cruzeiro do Sul). A controversial figure of this period present in both decrees mentioned (decree No. 21.415 from May 17, 1932, and decree No. 21.414, from May 17, 1932), Monteiro Lobato was object of numerous works that describe different versions and positions of him on the theme petroleum.

From 1932, in Federal decrees a series of Brazilian companies appeared: Companhia Nacional para Exploração de Petróleo (National Company for Oil Exploration—1932), Companhia Brasileira de Petróleo Cruzeiro do Sul (Brazilian Oil Company Cruzeiro do Sul—1933), Companhia Geral de Petróleo Pan-brasileira (Pan-Brazilian General Oil Companzy—1934), and Companhia Mato-grossense de petróleo (Mato Grosso Oil Company—1938). At that time, some societies were also created, such as the Sociedade Brasileira de Pesquisas Mineralógicas ltda. (Brazilian Society of Mineralogical Research Ltda.—1937), the Empresa Nacional de Investigações Geológicas Ltda. (National Geological Research Company Ltda), among others. From 1941, a few Federal decrees authorizing the creation of organizations or corporations were considered retroactive and their continuity was no longer authorized. With this federal intervention, oil exploration was moving toward state monopoly.

The path to state monopoly was closely related to the work done by the government initiative in the analyzed period. First, the discussion is centered again on topics that refer to the federal government, not to the state, as already mentioned elsewhere. However, it would be inappropriate not to direct a topic for the work carried out by the Geographical and Geological Commission of São Paulo, its role in the development of oil-related research and technology, which somehow, has repercussions on national research.

Thus, on March 27, 1886, the Geographical and Geological Commission of São Paulo (CGG) is created, "which owed its creation to the practical demands placed by the coffee industry of São Paulo" (Figueirôa 1997: 163). The work carried by CGG were aiming several fields: "Geology, Botany, Geography, Topography, Meteorology, Zoology, Archeology etc., in an attempt to produce a profile, as accurate as possible, of the São Paulo physical environment" (Figueirôa 1997: 167). The CGG also "started the gathering of the geographical, geological and topographical chart" of the province of São Paulo, "following the method called 'triangulation', adopted on that time by the '*U.S. Coast & Geodetic Survey*'" (Figueirôa 1997: 169). It had as first director the illustrious figure of Orville A. Derby.

In 1927, already under the direction of the engineer João Pedro Cardoso, the services of Exploration of the subsoil were created (Law No. 2.219, from December 9, 1927), that subdivide in service of apatite, that consisted of the research and the exploration of apatite deposits in the region of Ipanema (around Sorocaba City), and Petroleum service,[68] which aimed to meet the energy demand of Sao Paulo industry, a growing problem (Figueirôa 1999):

> The Petroleum service, in turn, carried out systematic researching of this energetic mineral, concluding as promising the geological picture of the State. The results of these studies were published in 1930 in Bulletin n. 22, Petroleum Geology of the State of São Paulo, Brazil, by Chester Washburne, a North American geologist with extensive experience in petroleum research in several regions of the world, brought specially to conduct this type of work with the commission (Figueirôa 1997: 122).

[68]For the first time, this service carried out a systematic search for oil in Sao Paulo territory. The results were later confirmed by the works of Petrobrás (in the 60s) and Paulipetro (in the 80s).

He was hired by the Sao Paulo government to study the oil problem in the State, Chester Wesley Washburne (1883–1971) presented, after 2 years of observations, his report, published in the abovementioned Bulletin No. 22, which defended the hypothesis of finding oil in São Paulo State, in "derived from Devonian rocks and suggested the drilling of several pointed structures. Just rehearsed a program based on their advice was soon abandoned before definitive conclusions" (Abreu 1948: 140).

In this period, "the Commission concentrated its efforts on oil drilling and on the detailing of the apatite from Ipanema deposits. This work continued until the last year of its existence." (Figueirôa 1997: 122). In spite of numerous political and economic problems, CGG "did not die and it was recovered, with modifications in 1935, with the creation of the Geographical and Geological Department, transformed in 1938 in the Geographical and Geological Institute" (Figueirôa 1997: 122).

Another important governmental initiative was the formation of the Commission of Studies of the Stone Coal Mines of Brazil. In 1904, Minister Lauro Müller (1883–1926),[69] invited the American geologist, Charles Israel (1848–1927),[70] to conduct the first coal exploration work in the region of Araranguá (Santa Catarina State), from where Criciúma City at the time was a district "(Klauck and Brunetto 2013: 99). The Commission for the Commission for Studies on Coal Mines was created by the Ministry of Industry, Transport and Public Works, in the face of the growing need for energy resources, led by White himself, who was "already an internationally recognized geologist, having been the first geologist to define the coal formation and reserves of Pennsylvania and Ohio" and in 1897 he was the first President of the West Virginia Geological and Economic Survey—United States (Orlandi Filho et al. 2006: 1). He was also a pioneer in applying the anticlinal theory[71] to locate wells in oil exploration:

> With the name linked to the anticlinal theory, which was providing fruitful guidance in the selection of areas for the discovery of oil discovery wells, White claims in the "Brazilian Engineering and Mining Review" that he was often asked because petroleum, already a reality in many countries, had not yet been discovered in Brazil (Moura and Carneiro 1976: 91).

The significant work of White and his collaborators—John H. Mac Gregor,[72] and David White (1862–1935),[73] supported by a team of Brazilian technicians and officials— executed between 1904 and 1906 in region of the Serra do Rio do Rastro, in Santa Catarina, became a major relevance for the stratigraphy of the Paraná Basin (Orlandi Filho et al. 2006). However, the great question of petroleum would appear in the publication of the final report of the White Commission in 1908, in which the geologist concluded, "in only two pages devoted to the petroleum potential on the area, that the possibilities are all against oil discovery, in commercial quantity, in any part of southern Brazil" (Marinho Júnior 1989: 217).

These conclusions were added after the drilling of "a well in Irati, in the state of Parana, showing the impossibility of occurrence of oil in that region due to the presence of volcanic rocks in the midst of sedimentary rocks" (Morais 2013: 40), except in the Devonian lands of the Amazon Basin.

In the report, White was frankly pessimistic about the petroleum potential of the region (Smith 1978), moving away "for some years from Brazilian cogitations the problem of liquid fuel research" (Pedreira 1927: 14). He was not the last to reject the claims about the existence of petroleum in Brazilian territory, as we will see in the course of this book.

As much as White's ideas had a strong impact in the period, they served as fuel for discussions

[69]Brazilian military, engineer, politician, and diplomat.

[70]North American geologist and professor.

[71]Designation of the theory that petroleum and natural gas migrate to the higher portions of the permeable layers and thus they are found in the anticlines.

[72]No biographic references were found.

[73]North American geologist. He made one of the most complete studies of Flora Glossopteris, the main fossil component of the mineral coal deposits in Brazil and Uruguay.

to boost oil research. Not agreeing with White's words, some geologists have taken initiatives, such as the civil mine engineer Miguel Arroyo Lisboa (1872–1932),[74] in his work O problema do combustível nacional (The problem of the national fuel), from 1916:

> the reasons (...) given by I. C. White, in 1906, for the abandonment of oil research in our land, in the south, are now ungrounded. On the contrary, it is precisely in the disturbed region of the eruptive rocks of the south, where they cut the Devonian terrain, rich in animal fossils, and also carboniferous, that we should search for oil (Arrojado Lisboa, M. apud Moura and Carneiro 1976: 93).

The above quote also points us another question: oil research based on paleontology studies. In 1948, Sylvio Fróes de Abreu reports that the conclusions reached by White were directly associated with Brazil "difficult geological reading":

> The great handicap in the south is the basaltic spills that cover most of the area with thick sheets, leaving only the border of the States of Sao Paulo, Parana and Santa Catarina, where it is easier to search for oil. Even in this range, the dikes, sills and lacolites of basic eruptions are frequent, cutting the sediments until the Triassic, which led I. C. White, in 1906, to consider as unlikely the existence of oil in southern Brazil (Abreu 1948: 138).

In 1908, new oil discoveries in Mexico, in an area severely cut by basic eruptions, the same as in southern Brazil, began to offer the country any chance of finding oil years later. This can be seen in the work of Eusebius de Oliveira, in the annals of the School of Mines in 1917, when referring to oil exploration in Mexico.

White's conclusions remained and somehow introduced a certain pessimism in finding oil in the south of the country, so much that the research was later concentrated in the State of São Paulo and especially in the northeast region of Brazil. If we look at the decrees in the period of publication of the White report (1908), we will observe that the only ones related—which appear 4 years later—refer to the establishment of foreign companies. With regard to private initiatives, they only come back from 1932, as we saw earlier, after the economic and political reforms already mentioned.

During this period, the CGG (already mentioned) and the SGMB (hereinafter described), both governmental initiatives, played a fundamental role in pursuing research and studies on petroleum.

In 1907, the Geological and Mineralogical Survey of Brazil (SGMB), under the leadership of Orville Derby, was installed in Rio de Janeiro, "an American scientist of the group of geologists who, from 1875, gave rise to the investigations of our Geology in the Geographical and Geological Commissions of the Empire, in Rio de Janeiro; and, after the Republic, in the State of Sao Paulo" (Vargas 1994: 214). The main objective of the SGMB was to obtain concrete and detailed information on geography, relief, geological structure, communication routes, mineral wealth, and soil types (Figueirôa 1997), but their work went beyond. Milton Vargas clarifies that, effectively, three institutions[75] actually started technological research in Brazil, one of them was the SGMB (Vargas 1994).

In 1908, the SGMB had already carried out "a broad research of iron and manganese reserves in Minas Gerais" (Figueirôa 1997: 223) and had a strong concern with energy issues. In addition to petroleum, the service began to be interested also in its derivatives, such as pyrobituminous shales and peat. As for energy issues, still in the management of Orville A. Derby, the investigations of petroleum prospects in the national territory began, and Euzébio Paulo de Oliveira was in charge. The results obtained by Oliveira until 1918 were published in 1920, in the first number of the SGMB Bulletin, *Brazilian Petroleum Rocks* (Figueirôa 1997: 226). With the death of Derby, Gonzaga Campos took over the leadership of the SGMB from 1915 until 1922. His

[74]Miguel Arroyo Ribeiro Lisboa graduated as a mining and civil engineer by School of Mines in 1894. He actively devoted himself to the exploitation of the country mineral resources and engaged in petroleum research in São Paulo.

[75]The other two institutions mentioned by Milton Vargas are: the Office of Mechanics of Materials of the Polytechnic School of Sao Paulo and the Agronomic Institute of Campinas.

activity, as a young man, was directed to the research of the coal mining area of Santa Catarina and to the petroleum in Maraú, Bahia State (a region also studied by Derby). Under the command of Gonzaga Campos, the SGMB also investigated:

> […] coal deposits, mineral occurrences of any kind that could be interesting to economic use and guided by similar motivations to the Geological and Geographical Commission of Sao Paulo, led to the gathering of the main sources of hydraulic energy in Brazil, producing profiles of waterfalls and studies of the river regime (Figueirôa 1997: 226).

Two divisions were also created in the SGMB, "which were geared specifically to the solution of national energy problems: the Hydraulic Forces service and the Experimental Fuel and Mineral Station" (Vargas 1994: 214). In addition, "it was with the installation of the Experimental Station in 1920 that the origin of the national technological research in the field of fuels and minerals" (Vargas 1994: 214). In that period, one of SGMB main concerns was related to the scientific aspects of geological issues and the energy issue.

Oliveira (2010) adds that the "Federal Government's interest in oil research in São Paulo came under the influence of Gonzaga de Campos […] it was impressed by the abundance of bituminous sandstones in the State, city of São Pedro" (Oliveira 2010, p. 36), specifically. "During the direction of Eusébio P. de Oliveira (until 1937), oil field research continued, albeit to a limited extent, given the precariousness of their equipment" (Figueirôa 1997: 227) and information related to probability of finding oil. Again, in 1934, the research of the Geological and Mineralogical Service of Brazil/National Department of Mineral Production condemned the south of the country through the argument presented in Bulletin No. 5 of the Mineral Production Promotion Survey, "Gondwanic Rocks and Petroleum Geology of Southern Brazil", by Victor Oppenheim (1906–2005),[76] and the negativist ideas of I.C. White (Abreu 1948). As we can observe:

> […] the existence of oilfields "in quantities and conditions industrially exploitable" in the south of the country was doubted and it was suggested that the research should be directed to the "little known geological territories of the extreme west and northwest of Brazil, in the areas bordering Bolivia and Peru", where "intense research and drillings must be carried out". Regarding the southern area of the country, however, reference was made to "bituminous horizon of Irati" with its bituminous shales that "present a great value as a national fuel reserve" and its exploration "could be performed by the usual extraction processes and distillation of bituminous weights" (Braga, Odilon apud Cohn 1968: 26–27).

Brazil was going through ups and downs when it comes to petroleum, due to lack of budget, technical staff, partly due to foreigners, lack of technical training or the simple—but relevant—reason for not having found a well of commercial or even subcommercial value. Despite this, the central question was still the training of the technical staff of the SGMB, formed by Brazilians and foreigners, and initiatives to send technicians abroad in the quest for new technologies, as it can be seen in the paper *O Paiz*: "The expedition was sent to Europe, in commission of the Ministry, of the student of the Geological Mineralogical Service, Theophilo Henry Lee" (Ministério da …, 1920, p. 3). Theophilo was an English chemist, hired under renewed contracts regimen almost every year.

This SGMB initiative would later reflect on the attitudes of the National Petroleum Council and Petrobras, which focused on the improvement and professionalization of professionals in the area of geosciences.

2.5 The Creation of the National Petroleum Council (1938) and the Discovery of the Petroleum in Brazil

The 30s, marked by economic and political changes, has brought significant advances in the steel industry—through a growing demand for steel coming from the process of urbanization (construction) and industrialization (Dantes and Santos 1994: 229)—in the progressive

[76]Victor Oppenheim was born in Latvia. Geologist.

nationalization of the mines, mineral deposits and waterfalls, in legislative reforms, and in the greater interest for liquid fuels, with particular attention to the petroleum.

The economist Pedro Paulo Zahluth Bastos adds that the construction of a nationalist policy in this period followed a similar dynamic for the steel and oil industry (Bastos 2006).

In 1937, the political establishment of the New State (*Estado Novo*) by Getúlio Vargas "transformed the relationship between federal authority and state authority, bringing Brazil closer to a truly national government" (Skidmore 2010: 65). However, political changes related to two Federal decrees would change this scenario in 1938: Decree-Law No. 395 from April 29, 1938, and Decree-Law No. 538 of July 7, 1938. Both will be discussed in detail in the second part of this book, but here we mention them in order to address a government initiative in the period covered.

According to Getúlio Vargas, the activities of CNP would serve a double purpose: industrialization and oil research. In addition, its inception served more to organize and secure the future discovery of oil:

> The Federal Government preferred a state solution in this first phase of CNP organization, without being monopolistic. It sought to improve a legal system capable of giving the State greater control over industry and, above all, neutralizing the sector of foreign capital pressures (Marinho Júnior 1989: 242).

CNP was directly subject to the orders of the President of the Republic, having been "conceived as an organ endowed with a high degree of administrative and financial autonomy" (Cohn 1968: 53). From then on, the responsibility for prospecting and exploration would be directed to the CNP.

At first, this Council faced difficulties with its staff of technicians as there were few people professionally trained to expand the industrial area of petroleum. In this way, its creation has become "a pioneering attitude related to the Petroleum Statute, regulation and more concrete initiatives of technical qualification" (Pinto 1988: 24).

In 1939, the discovery of oil in Brazilian soil, in the district of Lobato in Bahia State, boosted the development of the oil sector to a new level. In that year, CNP sent the first Brazilian technicians to be trained in the United States and Europe. But how did the discovery of petroleum happen?

In 1932, Bahia state surveyor, Manoel Ignácio Bastos, informed the President of the Republic, Getúlio Vargas, about the discovery of oil in the district of Lobato, Bahia State. The story says, known through newspapers of the period, that Manoel Bastos

> […] was digging a shallow cistern in his yard when he suddenly realized that the water he was searching was not so pure: before it began to flow a black oily liquid, which did not come as a surprise to the locals who had long ago already used it to feed their lamps. It was oil, almost under the skin (Em 1932 …, 1974: 3).

The evidences from Bastos were seen with a certain skepticism, for several times stories like his were told and nothing was found. In 1933, Manoel Bastos got in touch with Oscar Cordeiro,[77] defined as an "intelligent, astute and very well-connected man" (Em 1932 …, 1974: 3). Foreseeing a promising future around the possible discovery of oil, Cordeiro formed "society" with Bastos, to "soon to remove him definitively from the question, he made this by putting in Lobato a plate with the following words: 'Oil Mine belonging to Oscar Cordeiro' " (Em 1932 …, 1974: 3).

Then, according to Pedro de Moura, the adventure of publicity and the complete withdrawal of the real discoverer of the oil began. Cordeiro tried to prove the existence of oil in the district of Lobato countless times, but he was always overcome by official disinterest, translated into contrary technical opinions (Cohn 1968: 13).

Between 1933 and 1934, the American geologist and petroleum specialist Victor Oppenheim was hired by the Federal Government "to carry out studies in Alagoas, southern Brazil and Acre, expressing unfavorable opinions for the first two

[77]No bibliographic references were found.

regions" (Em 1932 …, 1974: 3). When he was in Bahia, Victor Oppenheim denied the existence of oil on the site, sending a telegram to the director of the SGMB, back then Euzébio de Oliveira, claiming that Lobato oil was a "mystification". Oppenheim's unfavorable opinion reinforced the view adopted by the government (Em 1932…, 1974).

This "mystification" ravaged Brazil for years. Meanwhile, Oscar Cordeiro continued to affirm the existence of oil in Lobato. Several figures entered the scene, some denying the oil and others stating their existence. Today, Victor Oppenheim's statement is made clear by the geological flaw at the site—which he did not diagnose—that juxtaposes metamorphic (unfit for oil) rocks and sedimentary rocks, where there might be oil:

> The oil exudation in Lobato, Bahia was a geological puzzle. Because of the following: in a cacimba (cistern), on the beach, there was oil, and a hundred or two hundred meters away there were metamorphic formations where oil could not exist. It was a flaw. And there was no knowledge of local geology, so that some important geologists thought that it was the remains of oil from old fuel deposits from the construction of the port of Bahia (Pinto 1988: 24).

As already mentioned, the difficulty in observing geologically the Brazilian territory through comparative studies with other places caused these types of technical errors, such as Oppenheim's.

In 1935, some flasks containing Lobato oil samples were shipped to the National Institute of Technology in Bahia. Hope was rekindled, the engineer Sylvio Fróes de Abreu wrote: "The conclusion we draw from the study of the samples in the laboratory and the observation on the ground is that seems to be a clear indication of oil in the Cretaceous Todos os Santos Basin" (Pinto 1988: 136–137). In 1936, Sylvio Fróes de Abreu published the work Contribution to the Geology of Oil in the Recôncavo, in collaboration with Glycon de Paiva and Irnack do Amaral:

> This work was essential for the discovery of oil in Bahia. His opinion, however, was that Brazil was a country with no energy, since its resources in coal and oil were modest, and bituminous shale, which here occurred in large numbers, was of poor quality. It proposed the technological development in the direction of energy use in our country. In the meantime, every effort was made to publicize the existence and viability of oil exploration in the Recôncavo Baiano (Vargas 1964: 217).

In 1937, initiatives such as the one made by Avelino Inácio de Oliveira[78] gave a new configuration to the problem mentioned above, when, through the DNPM, they sent a small drill to verify the site, which had 600 m of capacity and that left the deposit from Ponta Grossa City going to Salvador City (Maya 1938). The works of "geological mapping of the Recôncavo Baiano were conducted by the geologist Pedro de Moura, and the geophysics was in charge of a United States Company, the United Geophysical Co." (Dias and Quaglino 1993: 22), until the discovery of petroleum in January 21, 1939 (Paiva, G. apud Oliveira 1940: VIII):

> the government immediately issued a decree instituting a national reserve within a 60-kilometer radius, enough to cover the areas already granted to private research in Candeias, Itaparica, Montenegro, Matarandica, Santo Amaro, and to eliminate the private initiative throughout the Recôncavo Baiano (Abreu 1948, p. 140).

The technological activity focused on the study of petroleum between 1864 and 1939 found barriers in the use of technical methods and problems, which were focusing on the production of the necessary industrial products and in the use of imported technology:

> the brochure from Morais Rêgo: Possibilities of the existence of Oil in Bahia (1932) is the first technical publication releasing this idea safely; the book Contributions to Oil Geology in the Recôncavo (1936), from Abreu, Paiva and Amaral, is the first work with positive data on the issue; the 1937 geophysical studies consolidated the idea even more, and finally the 1938 hole definitively proved the existence of oil in the Recôncavo. In January 1939, a new era of oil in Brazil begins (Abreu 1948, p. 140).

[78]Graduated in Mining and Civil Engineering by the EMOP in 1916. Served as Director from the DNPM from 1951 to 1961.

OURO NEGRO NO BRASIL!

Confirmada, finalmente, a existencia de petroleo nos terrenos de marinha da região de Lobato

ULTIMA EDIÇÃO 17 HORAS

Vae á cidade do Salvador o geologo Glycon de Paiva-As providencias do ministro da Agricultura – Fala ao GLOBO o general Horta Barbosa, presidente do Conselho N. do Petroleo

O GLOBO

NOVOS RUMOS PARA A ECONOMIA BRASILEIRA!

CHEGA HOJE o Ministro da Guerra

Fig. 2.4 Black gold in Brazil (1939). *Source* Ouro Negro …, 1939

The drillings were concentrated in the district of Lobato and after 17 wells—seven with petroleum, ten dry and 4 years of activity, the he verdict was incontestable: subcommercial (Pinto 1988) (Fig. 2.4).

Only in 1941 that the first commercial oil field in Brazil was discovered in Candeias (Bahia State). The discovery of oil served as an impetus for the industry and for the activities that relied on it, stimulating later the creation of geology courses and specific training in Brazil, as well as the creation of Petrobras, as we will see next.

References

Barbuy H (1999) Exposição universal de 1889 em Paris. Loyola, São Paulo

Bastos PPZ (2006) A dinâmica do nacionalismo varguista: o caso de empresas estatais e filiais estrangeiras no ramo de energia elétrica. In: Encontro Nacional de Economia. Anais …. Associação, Salvador

Bongiovanni LA (1994) Estado, Burocracia e Mineração no Brasil (1930–1945). Thesis (Master Degree in Geosciencie), University of Campinas

de Carvalho JM (2010) A escola de Minas de Ouro Preto: o peso da glória. Centro Edelstein de Pesquisas Sociais, Rio de Janeiro

Chiaradia K (2008) Ao amigo Franckie, do seu Lobato: estudo da correspondência entre Monteiro lobato e Charles Franckie (1934–37) e sua presença em “O Escândalo do Petróleo” (1936) e “o Poço do visconde” (1937). Thesis, Master Degree in Theory and History, University of Campinas

Cohn G (1968) Petróleo e Nacionalismo. Difusão Européia do livro, São Paulo

Dantes MAM, Santos JS (1994) Siderurgia e Tecnologia (1918–1964). In: Motoyama, Shozo (org) Tecnologia e Industrialização no Brasil—Uma perspectiva histórica. UNESP, São Paulo, pp 209–250

Dias JL de M, Quaglino MA (1993) A questão do petróleo no Brasil: uma história da Petrobrás. Fundação Getúlio Vargas, Rio de Janeiro

Figueirôa SF de M (1997) As Ciências Geológicas no Brasil: Uma História social e institucional, 1875–1934. Hucitec, São Paulo

Figueirôa SF de M (1999) Ciências, elites e modernização em são Paulo (1886–1931). In: Ferreira AC, Luca TR, Ioki Z (Org) Encontros com a História: percursos históricos e historiográficos em São Paulo. UNESP, São Paulo, pp 113–129

Freitas Filho AP (1991) Tecnologia e Escravidão no Brasil: aspectos da Modernização agrícola nas Exposições nacionais da segunda Metade do século XIX (1861–1881). Revista Brasileira de História 11 (22):71–92. São Paulo

Gusmão LG de S (2005) Recursos Enegéticos. In: Serafim CFS, Chaves P de T (org) O Mar no Espaço Geográfico Brasileiro. Ministério da Educação; Secretaria de Educação Básica, Brasília

Heizer A (2009) Ciência para todos: a exposição de Paris de 1889 em revista. Revista de História e Estudos Culturais, ano 6 6(3)

Hobsbawn E (1985) Nação e Nacionalismo. Revista Ler História (5). Lisboa

Klauck AG, Brunetto S (2013) O Mapa da Minas. Revista Santa Catarina em História 7(1). Florianópolis

Leoncy LF (1997) O regime jurídico da mineração no Brasil. Paper 073 do NAEA. http://www.ufpa.br/naea/novosite/paper/117. Accessed Aug 2013

Mansano RB (2004) Engenharia de Perfuração e completação em poços de petróleo. Federal University of Santa Catarina. Lecture, Florianópolis

Marinho Júnior IP (1989) Petróleo: política e poder: um novo choque do petróleo? José Olympio, Rio de Janeiro

Martins L (1976) Pouvoir et développement économique —formation et evolution des structures politiques au Brésil. Éditions anthropos, Paris

Miranda FCP de (1960) Comentários à Constituição de 1946. 3. ed. rev. e ampl. Borsoi, Rio de Janeiro

Monteiro DM (2002) Política de terras no Brasil: elite agrária e reações à legislação fundiária na passagem do império para a república. Revista História econômica & história de empresas 2:53–73

de Morais JM (2013) Petróleo em águas profundas—Uma história tecnológica da Petrobras na exploração e produção offshore. Instituto de Pesquisa Econômica Aplicada e Petrobras, Brasília

Moura P de, Carneiro FO (1976) Em busca do petróleo brasileiro. Fundação Gorceix, Ouro Preto

Oliveira JCT de (2010) A História do Petróleo no Estado de São Paulo, antes do monopólio da Petrobras (1872–1953) Final Work, Geology Degree, Campinas: University of Campinas, 78 f

Oliveira LL (1990) A Questão Nacional na Primeira República. Brasiliense, São Paulo

Orlandi Filho V, Krebs ASJ, Giffoni LE (2006) Coluna White na serra do rio do rastro, SC; Seção-Tipo de Unidades do Continente Gonduana no Brasil. In: Winge M et al (ed) Sítios Geológicos e Paleontológicos do Brasil. http://www.unb.br/ig/sigep/sitio024/sitio024.pdf. Accessed Apr 2014

Peixoto JB, Peixoto W (1957) Produção, transporte e energia no Brasil [s.l.: s.n.]

Peyerl D (2010) A trajetória do paleontólogo Frederico Waldemar Lange (1911–1988) e a História das ciências. Thesis, State Ponta Grossa University, Ponta Grossa, 116 f

Skidmore TE (2010) Brasil: de Getúlio a Castello (1930–64). Companhia das Letras, São Paulo

Smith PS (1978) Petróleo e política no Brasil Moderno. Artenova, Rio de Janeiro

Solberg CE (1979) Oil and nationalism in Argentina: a history. Stanford University Press, Stanford

Souza PJB de, Lima VL (2002) Avaliação das técnicas de disposição de rejeitos da perfuração terrestre de poços de petróleo. Final work, Specialization in Management and Environmental Technologies, Federal University of Bahia, Salvador City

Tarbell IM (2003) The history of the standard oil company. Dover Publications

Távora J (1955) Petróleo para o Brasil. José Olympio: Rio de Janeiro

Thom WT Jr (1929) Petroleum and coal—the keys to the future. Princeton University Press, Princeton

Vargas G (1964) A política nacionalista do petróleo no Brasil. Tempo brasileiro, Rio de Janeiro

Vargas M (1994) O início da pesquisa tecnológica no Brasil. In: Vargas M (org) História da técnica e da tecnologia no Brasil. Editora da Universidade Estadual Paulista, Centro Estadual de Educação Tecnológica Paula Souza, São Paulo, pp 211–224

Yergin D (2010) O petróleo—Uma História de Ganância, dinheiro e Poder. Paz e Terra, São Paulo

3 The Formation of the Know-How (1938–1961)

3.1 The Brazilian and Foreign Work for the Oil Industry Formation in Brazil

In 1937, in Argentina, the internal consumption has already been achieved due to the activities of the state Company *Yacimientos Petroliferos Fiscais*; in Uruguay the refinery Teja was opened; in Mexico, the famous oil crisis with the United States took place, and in Bolivia, 1 year before, a state company was created to expropriate the *Standard Oil Company* from the territory (Martins 1976: 288). At the same time of those events, in Brazil, there was still the discussion about the existence of oil, concentrating the efforts from technicians and entrepreneurs in the control of the oil exploration and refining (extremely poor and with an oligopoly that controlled the import and distribution of fuel) (Martins 1976: 288–289).

To Gabriel Cohn, in his work *Petroleum and Nationalism*, from 1968, in the end of the 1930s, Brazil was living the policy of the New State (1937–1945); in the international panorama, the beginning of the World War II (1939–1945) was worrying the military from Vargas Government; and in the national economic plan, it could be observed an industrial and expansion impulse of the road network. They are considered by the concerned author as boosters of the public sector in actions taking to solve problems that involved the petroleum.

It is possible to add other two factors that contributed to such attitude: (a) the expropriation of all foreign companies and the nationalization of the Mexican oil in 1938 (significant influence not only in Brazil, but in Latin America as a whole); and (b) the repercussion of the publication of The Petroleum Memorial and National Defense (1936) by General Júlio Horta Barbosa (1881–1965)[1] for the Minister of War and General Eurico Gaspar Dutra (1883–1974)[2] in which he calls attention to the oil problem and criticizes the "timid explorations" for oil obtainment, stating that

> without the oil, our military potential is low; without the oil, we will watch sadly the constant and uninterrupted penetration of the standard oil, of the Royal Dutch Shell, Mexican Éagle, through the smallest corners of our homeland. Today, we will replace these names by Brazilian names. We are deeply hopeful that such a substitution will occur sooner as the effectiveness of the army's cooperation in locating the sources of such fuel (A questão … 1979: 3).

[1]Brazilian Military and "Sertanista". He graduated in Engineering and has a bachelor's degree in Mathematics and Physical Sciences. He presided the Military Club between July 1936 and January 1937, when he got involved in the debate about the existence or not of petroleum in the Brazilian subsoil. He was president of the CNP from 1938 to 1943. He became a proponent of the state oil monopoly and joined the Petroleum Campaign. Horta Barbosa always participated in the fight of the nationalist sectors that allowed the creation, in 1953, of Petrobras.

[2]Brazilian military and 16th President of Brazil. He actively participated in the establishment of the New State (*Estado novo*). He won the elections for the presidency of the republic in 1945, assuming the government on January 31, 1946, leaving the government on January 31, 1951.

D. Peyerl, *The Oil of Brazil*, Historical Geography and Geosciences,
https://doi.org/10.1007/978-3-030-13884-4_3

The Brazilian government was then pressured to create a separate agency for the petroleum issue, as the low operating conditions of private and government initiatives persisted, highlighting the "impossibility of few technicians specializing abroad, […] equipment in discomfort with the operational needs, […] and the hostile geology of the sedimentary basins" (Moura and Carneiro 1976: 225).

On April 29, 1938, the promulgation of Decree-Law No. 395, which declares "of public utility and regulates the import, export, transportation, distribution and trade of crude oil and its derivatives in the national territory, and thus the refining industry of imported oil or produced in the country" (Cohn 1968: 50). In its article No. 4, this normative act created the National Petroleum Council (CNP). It is then the first attitude of nationalization of the oil refining industry in the country. Still, in the preamble of this Decree-Law, it is emphasized:

> […] that these measures are taken "considering that petroleum constitutes the main source of energy for the accomplishment of transportation, especially air and road transport, a national public utility, indispensable to the military and economic defense of the country" and "considering the coexistence of economic order to provide for the distribution throughout the national territory of oil and oil products at prices as uniform as possible" (Cohn 1968: 50).

Right after that, on July 07, 1938, the Decree-Law no. 538 organizes the then established CNP, defining its attributions and dividing it into "a deliberative agency (the Full Council) and technical and administrative agencies" (Cohn 1968: 58). Thus, it was only in charge of the CNP the responsibility for controlling and authorizing the research, exploration, import, export, transportation, distribution and trade of oil and oil products. Regarding the refineries, it was the responsibility of the CNP to authorize the installation (only for Brazilians) and also to supervise the operation of any of them, that is, the organ assumed "all the responsibilities of the oil industry, outlining what would be an oil policy in the country" (A questão … 1979: 3).

This period, which starts in the promulgation until the execution of the Decree-Laws referred, is very well defined in the work Searching for the Brazilian Oil (*Em busca do petróleo brasileiro*), from 1976, by Pedro de Moura and Felisberto Carneiro, as being a turning point, as from this moment, the oil policy has had administrative and financial autonomy to go its own way. This autonomy was established by Decree-Law No. 1.143 from March 9, 1939 (Brasil 1939) and was visible when compared to the situation of flexibility in the management of funds between the CNP and the National Department of Mineral Production, the CNP's being much larger (Dias and Quaglino 1993: 14).

The beginning of work by CNP was surrounded by several problems already known and accumulated during the various attempts of oil exploration since the end of the nineteenth century, as pointed out in the first part of this book. The main technical difficulties found were centered in four items:

(a) Extension and complexity of the territory to be explored, aggravated by the distance to the main urban centers of the country, lack of communication routes, lack of local resources, aggravated, sometimes by the existence of tropical endemics.
(b) Precariousness of geographic and geological maps, already available, of the country.
(c) almost complete ignorance of the structural (tectonic) conditions of the subsoil to be explored.
(d) Generalized lack of technicians and manpower (Távora 1955: 73–74).

The partial contribution to the solution of items *a*, *b*, and *c* appeared with the elaboration of a map that described, in gradation, the areas with possibilities finding petroleum in Brazil, published in 1938, by Avelino Ignácio de Oliveira, according to the geological knowledge of the time (Abreu 1948: 137). This map served as a base for the beginning of the CNP research. In addition, through the flexibility of funds earmarked for the aforementioned CNP, it was allowed to:

> [...] for example, the contracting of service-providing firms such as the North American Drilling and Exploration Co. in charge of drilling the wells, and United Geophysical Co. having no technicians or equipment in desired number, CNP was able, through these contracts, to accelerate the pace of work. It was even possible to acquire three modern probes in 1940. However, when restrictions on the shipment of exploration equipment and spare parts were enforced by the US government, there was no alternative but to slow the pace of the activities (Dias and Quaglino 1993: 14).

Both external and internal events affected CNP's pace of research. In an attempt to alleviate the difficulty described in item *d*, one of the measures taken was to transfer DNPM technicians and equipment to the CNP, which was made by Decree No. 1.369 from July 23, 1939 (Dias and Quaglino 1993). Such an attitude was connected to the orders of the first president of the CNP,[3] appointed by the President of the Republic, Júlio Horta Barbosa, who remained in charge until 1943. Horta Barbosa's policy, known for his nationalist attitude, was reflected in three distinct plans: (a) the elimination of private wild cat activities; (b) regulation of the fuel market; and (c) the creation of state-owned refineries (Martins 1976: 301) (Fig. 3.1).

Another reason that led CNP to take initiatives regarding the lack of manpower or equipment was the discovery, on January 21, 1939, of oil in the district of Lobato, Bahia State. So much that,

> between 1940 and 1942, Standard Oil made three proposals for the creation of joint companies for research and extraction, being rejected by the opposition of the military summit, despite the majority favorable position of the cabinet of Vargas ministers (Bastos 2006: 8).

In April 1939, 2 months before the CNP began to operate, Horta Barbosa focused on visiting the oil industries of Uruguay and Argentina the trip took him to

> [...] to recommend that as many technicians as possible are sent to the "Plata" countries to study the specialties of the oil industry; that CNP controls all activities related to the industry; that counselors are requested from Argentina; that refining became a state monopoly; and that more funds are allocated to CNP.[4]

In addition, the CNP, under Horta Barbosa command, ordered from the United States "a full set of machinery" (Moura and Carneiro 1976, p. 230), investing in the modernization of equipment ("acquisition of three rotary drills, i.e., two steam, with a capacity of 2500 m, and one diesel, for 1800 m" (Moura and Carneiro 1976: 230), which are destined for the states of Alagoas and Bahia. "North American experts trained Brazilians in the operation of the machines, who would work twenty-four hours a day" (Smith 1978: 54).

CNP also announced that would hire foreign companies specialized in drilling, which began in 1940 through the contracting of the *Drilling and Exploration Co.* (*Drillexco*), from Los Angeles (United States), which had as proposal/obligation "to train national coaches who were port of their teams" (Moura and Carneiro 1976: 230). It is noteworthy that not all technicians who participated in the training remained in the CNP.

CNP has also invested in foreign geophysical companies for seismic prospecting, contracting, for example, the *United Geophysical Co. S.A.*, from Pasadena, California (EUA). The services of these companies went beyond local research, also contributing to the training of Brazilians participants (Moura and Carneiro 1976: 230).

In that period, investment in refineries, as Horta Barbosa observed on a visit to the refineries in Uruguay and Argentina, was of utmost importance, as the potentially high profit from refining could finance the development of the oil research. For this reason, the government, as mentioned by Pedro de Moura and Felisberto Carneiro, "dated" the steel industry by concentrating its attention to

[3]In order to assume the position of president of the CNP, it was necessary: (a) to be a born Brazilian, of well-known competence and of an unblemished reputation, besides being over 30 years old; (b) be in the enjoyment of their civil and political rights; (c) have not, at the time of designation, had in the preceding 5 years direct or indirect interests in private companies engaged in or engaged in the research, mining, industrialization or trade of petroleum and its by-products (Brasil 1938).

[4]Information from the Câmara dos Deputados, 1939 apud Smith (1978: 53).

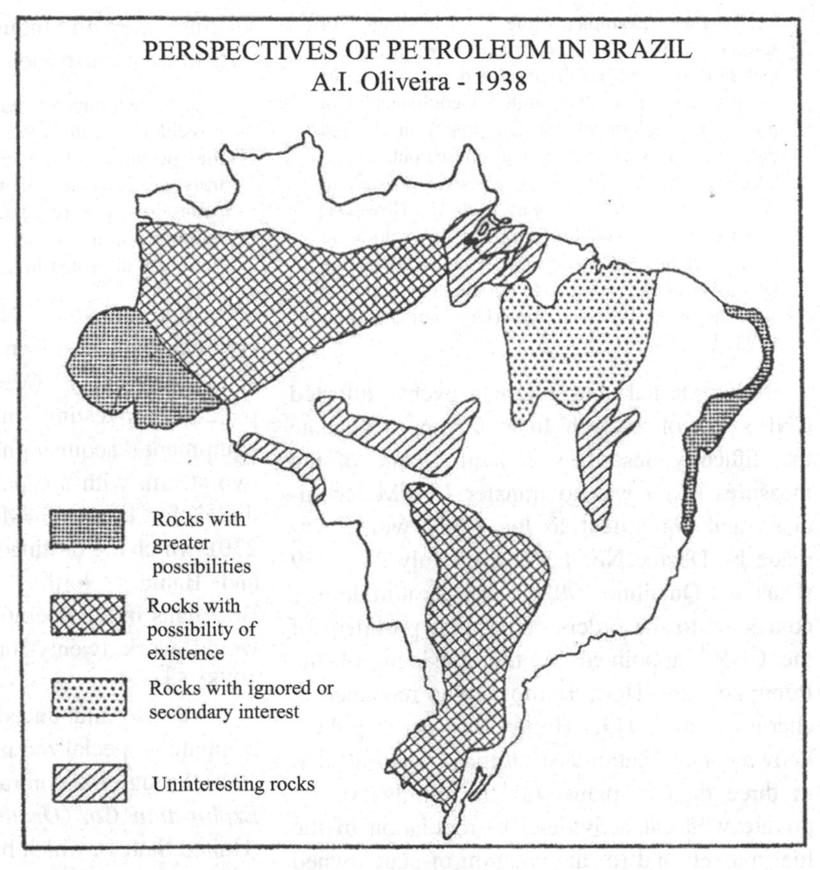

Fig. 3.1 Perspectives of petroleum in Brasil (1938). *Source* Abreu (1948: 137)

the construction of a steel mill (Moura and Carneiro 1976; Smith 1978; Cohn 1968). The solution found was to work so that the oil refining and research industry could be financed from within the oil industry itself (Smith 1978).

We clearly observed that the nationalization of the oil led to the exclusive privilege of exploring oil wells and to the profits acquired by the refining industry, not forgetting that the presence of foreigners was essential for the construction of the Brazilian know-how despite the laws and initiatives created to solve the oil-related issues since the creation of the CNP, the first official report of 1944 (published in 1946) mentions the same problems faced from the beginning by the agency: operating "during its early years with lack or inadequacy of equipment; with insufficient staff, […] and there are several examples of waste of time, equipment, and money" (Cohn 1968: 59). The criticisms that had been going through the years since the creation of the CNP were mostly related to the refusal to allow foreign capitals to participate in oil drilling and exploration, i.e. Brazil could contract and keep control of the situation in its hands, but not open to foreign investments.

Quickly, the CNP made "both the use of foreign experience and allowing nationalist sentiment," which was positive for the proper petroleum research (Smith 1978: 55). However, these criticisms were far from over, especially when discussions went back to the establishment of foreign companies or to the creation of a mixed foreign and national capital companies to carry out the work of drilling, production, and supply of technicians (Smith 1978).

The period of regulation and decisions presented by the CNP, mainly from 1939 to 1943, makes clear the oscillations of opening the participation and/or contracting foreign companies, such as the previously mentioned *United Geophysical Co.* and *Drilling and Exploration Co.*—in a country that had nationalism as its bias, but that needed the know-how of other countries to advance in the exploration and exploration of oil, according to Smith, "for many, it must have seemed unbelievable that a country so hopelessly in need of fuel would so vigorously reject foreign specialized technical assistance that it could accelerate the discovery of national sources of oil" (Smith 1978: 60).

This is what the new President of CNP, in 1943, tries to change. Army Colonel João Carlos Barreto (1895–1970) "was more interested in the rapid development of oil than in maintaining nationalist controls on the nascent industry," and was openly receptive to the idea of foreign participation in the discovery of oil (Smith 1978: 61)—so much, the added in the Report from 1944.

We were very concerned about the formation of a group of Brazilian technicians specialized not only in the petroleum geology, but also in all branches of exploitation and exploration.

From there, it was established that in all contacts with foreign companies or technicians for the execution of specialized services should be an essential and imperative condition for the formation of our people as soon as possible, that is why in the future would be up to him the responsibility of the technical conduction of such a vital problem in the Country. Concomitantly, we are looking to recruit more Brazilian professionals to the Board to work alongside the contracted North American technicians, in order to make the desired selection more feasible (Moura and Carneiro 1976: 244).

External factors would still change CNP initiatives, since much of the material needed for fieldwork and for the beginning of explorations was difficult to obtain abroad due to World War II (1939–1945).

However, we add to this fact that even in the most critical period of the war (1941–1942), "the participation in the industrial sector in the global product did not stop growing," and, after the conflict, it presented a remarkable advance (Cohn 1968: 74). Brazil also took advantage of this period to hire technicians and geologists, as well as to "put to the side of each foreign geologist [hired] two national technicians to train" (Moura and Carneiro 1976: 230).

Another event, internally driving in this study occurs from 1945, when investments for practical training within the CNP, in several areas, gain relevant attention directly related to petroleum exploration activities. These activities focus on drillers, blade operators, heavy truck drivers and boat pilots, dynamite load handlers, aerial photo interpreters, designers, mechanics, seismograph operators, among other types of petroleum research workforce (Moura and Carneiro 1976: 245).

To meet this need for manpower, CNP invested more and more in hiring foreign professionals to respond to the training of Brazilians. Foreign companies hired to explore Brazilian soil, in addition to receiving their services, still required to the CNP good conditions for carrying out the work, such as "access roads, encampments, bridges, land movement, admission of personnel to the most diverse professions, adjustments of contractors etc." (Moura and Carneiro 1976: 249). If these conditions were not possible, the foreign company would take the lead and make the necessary improvements, but the CNP would have to bear any cost beyond what was already established. Foreign companies that settled here together with the CNP moved about 1.500 employees (Moura and Carneiro 1976).

Most of the foreign companies that have settled under contract with CNP trained engineers in specific areas such as drilling. Meanwhile, other Brazilians from the CNP became specialized abroad, mainly in the United States, in petroleum engineering and geology.

Between 1945 and 1950, Brazil underwent a period of intense political revision of economic nationalism, joining the hiring of a greater number of "foreign technicians to help both in the elaboration of guidelines and in the training of Brazilians" (Smith 1978: 60). However, these training had still a purely practical nature, not an improvement or certified professionalization.

In the mid-40s, Sylvio Fróes Abreu pointed out that "the isolation of Brazilian technicians would act as a brake on Brazilian petroleum research and mining, and the lack of contact with foreign geologists had resulted in the Brazilian methods being behind" (Smith 1978: 64). Although there was already a position of the CNP in the face of this situation—because it was seeking foreign technicians to train Brazilians or sending part of the Brazilian technicians to specialize abroad—this was not enough to calm the moods of nationalists and *entreguista*[5,6] waiting for news of producers of oil wells.

> The National Petroleum Council was revealed as faulty in day-to-day activities, affected by the technical sector and the current administration, its importance was best demonstrated in the taking of the "big decisions", for which it had in fact been conceived rather than for routine action.
> It is, therefore, in the area of decisions that affect the economy of the country in a global way and refer to the main lines of oil policy that the CNP action has left its deepest marks (Cohn 1968: 61).

In 1947 Sylvio Fróes de Abreu sketched a map based on that of Avelino Ignácio de Oliveira, 1938, which he calls "Prospects of Oil in Brazil", contributing to the study and research of oil in the country. The contributions of Fróes de Abreu to the Oliveira map consist in the "increased geological conditions of new knowledge" acquired in the period, adding the rapid evolution of methods related to oil research worldwide, as well as the introduction of the "concept of accessibility and distance to consumption centers" (Cohn 1968: 136).

> That was very well synthetized by A. I. Levorsen,[7] when he organized an inquiry in 1942 to consult the opinion of technicians about the best way to discover oil and gas in the United States. Commenting on the fact, he highlighted the evolution of geologists' thinking; in 1920, he says, the answer would undoubtedly be: more mapping of surface structures; in 1925 would be: more emphasis on subsurface conditions, and in 1930 would be more geophysics. Now we add that in 1947 would be: more aerogeology, more sedimentology and more geochemistry (Abreu 1948: 136).

Regarding the concept of accessibility and distance to the centers of consumption on the map drawn up by Fróes de Abreu, the fundamental importance of the transportation of oil is highlighted. This is an essentially economic problem in the treatment of commercial and sub-commercial wells.

> It should be noted that a barrel of Brazilian oil in the confines of Acre is less valuable than a barrel in the Recôncavo or Paraná basin. Another factor to be considered is the accessibility of the area to the research, this depends partly on the distance to the civilized centers and also on the vegetation cover, salubrity, climate of the region and the geological conditions particular to the area. Five hundred meters of basalt over an area constitute almost an insurmountable obstacle to the techniques in use. Those who have made explorations of this kind know how they weigh the expenses of the aggression of the environment or isolation. The need for a well-stocked warehouse, food expenses, sanitary measures and transport raises so much the price of the work that the geographic conditions of the area begin to influence in a very sensible way on the cost of the barrel produced (Abreu 1948: 136).

In the late 40s and early 50s, changes in the political scenario changed the country economy. Dutra government adopted an open economic policy for multinational corporations, establish-

---

[5]Oil-related issues that should be resolved by attracting foreign capital.

[6]Juarez Távora, political and military, was known as a *entreguista* in relation to oil exploration in Brazil. He was in favor of the delivery of natural resources for the exploration of foreign companies and institutions, opposing the nationalist issues of the time. He was the main leader of those who opposed to the creation of Petrobras.

[7]Arville Irving Levorsen (1864–1965) was an American geologist.

ing an international political scenario and significantly increasing the number of imports. For the economist Pedro Paulo Zahluth Bastos, the economic policy of Dutra government can be described as a pendulum because:

> Initially, priority was given to an inflationary stabilization plan (limiting public spending, credit control) supported by liberal reforms (trade and financial openness) and nominal exchange rate stability. The diagnosis of inflation blamed interventionist inheritances of the New State, so that controlling prices required liberating market mechanisms and limiting harmful influences on government. Approximately halfway through the mandate, a currency crisis forced the government to turn back trade liberalization to defend foreign exchange reserves and safeguard essential imports, while not reversing the initial financial opening. The government tried to defend the exchange rate by limiting imports that are not essential, avoiding the inflationary impact of a currency depreciation. This protected the domestic market for substitute import production, which followed its course as the government sought to remove "bottlenecks" (shortages of debt, credit, and infrastructure) that limited the expansion (Bastos 2004: 100).

It was also during Dutra government that the Constitution of 1946 was enacted, when the social interests in determining the form of the state and directing industrialization in the country became clearer (Draibe 2004). The article 153 of the Constitution states that "the exploitation of mineral resources and hydraulic energy depends on federal authorization or concession in the form of the law" (Brasil 1946). The first paragraph of that article provides (Fig. 3.2):

> The authorizations or concessions will be granted exclusively to Brazilians or to companies organized in the country, and the landowner will be assured of a preference for exploitation. The landowner's preference rights to mines and deposits will be regulated according to their nature (Brasil 1946).[8]

To this article, it was summed to the ideologies constitutionally adopted in the Constitutions of 1934 and 1937 (Coelho 2009).

In 1947, through a commission created by President Eurico Gaspar Dutra, the objective was to revise the existing laws regarding the exploration of oil in the country and to create others, such as the bill entitled Petroleum Statute. The project sparked the debate between nationalists and *entreguistas*. If the project were to be carried out, the nationalization of oil would become impossible, since it would favor the opening to the international capital. Associated with this project it was the CNP decision "to open the oil refining industry to private companies" (Cohn 1968: 106). One of the main justifications for opening to foreign capital was the lack of specialized technicians and the precarious conditions in which CNP was undergoing (Cohn 1968; Smith 1978).

Reacting to this bill, nationalists gathered at conferences at the Military Club launch the Petroleum Campaign, with the slogan "The petroleum is ours". This campaign, which mobilized the military, intellectuals, students, the press, trade unions and the National Congress, aimed at national control over the petroleum, advocated state monopoly and raised questions about oil exploration and foreign influence on the product. "The speech identified with the *nationalist developmentalist* movement of the Military Club was so compelling that, in alliance with civilians, raised broad sectors of society in defense of the petroleum" (Andrade 1999, p. 81).

One point to emphasize in relation to the Petroleum Campaign is that, at the moment, there was no longer any discussion about the existence of oil, a matter that has gone through decades of controversy and conflicts. Until then, few wells (and still in low commercial quantities) were

[8] "The preference is between people who have the right capacity that the art. 153, 1st part (Brazilian or society, Brazilian legal entity, organized in Brazil). A company organized in Brazil but linked to foreign law (e.g. are legal entities of foreign law, or are subsidiaries or branches of foreign companies) cannot have mine or explore it, because the *ratio legis* of art. 152, part 1, of the 1946 Constitution was to remove all possible intervention from external action. The capital may be foreign; the society must be organized in Brazil and, therefore, only governed by Brazilian law. [...]. The actions of mining companies, subsoil wealth, and water supplies could not be to the bearer. In a collective name, limited partnership, none of the partners could be a foreigner, nor a society composed of foreigners, although Brazilian. Society, still Brazilian, so that it could be shareholder of some of these companies, had to be composed by Brazilians and only by Brazilians. no partner's right was alienable to foreigners" (Miranda 1960: 524–525).

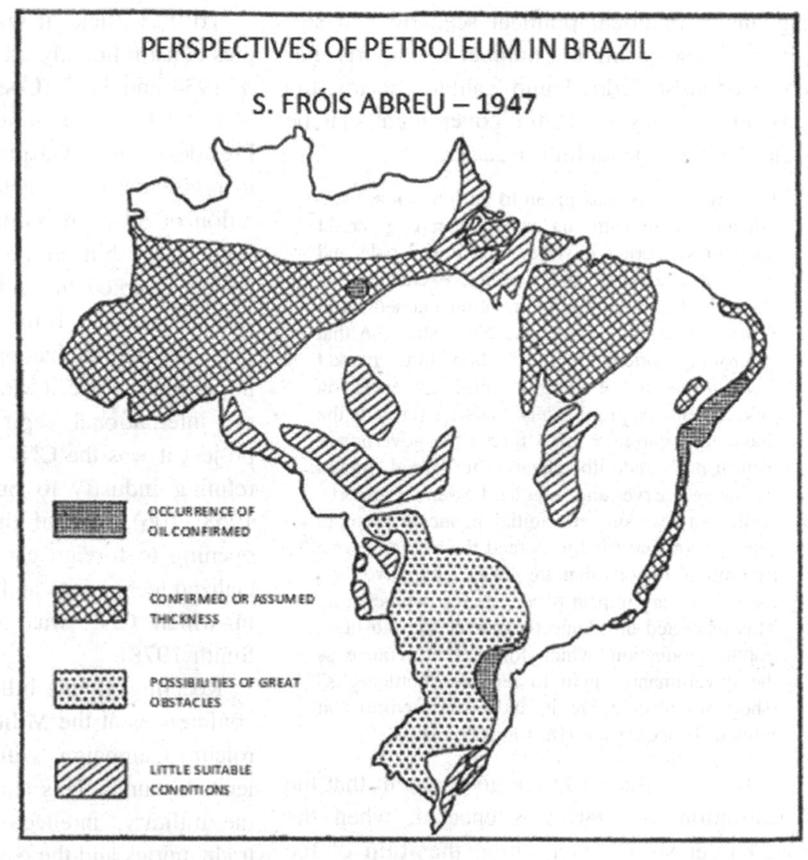

Fig. 3.2 Perspective of finding petroleum in Brazil—1947. *Source* Abreu (1948: 137)

being explored in the country. "'The Petroleum is ours' was only a generalized yearning for unrestricted nationalization of the oil industry in the country" (Moura and Carneiro 1976: 261).

After 2 months of the presentation of the Petroleum Statute in the National Congress, the Petroleum Studies and Defense Center was created in April 1948,[9] which began to govern the Petroleum Campaign in the country, "articulating military, students, public men and intellectuals" (Petrobras 50…, s.d.). The central purpose was to promote debates, conferences, articles, among others, turning to the nationalist strengthening of the state monopoly.

After the difficult process of the Petroleum Statute project, it is filed. Dutra government began to invest in the creation of refineries and the acquisition of a national fleet. In 1950, the National Tanker Fleet (FRONAPE) was created with the objective of "carrying out the transport of petroleum and oil products in Brazil and abroad, and may also carry out the respective storage and trade" (Brasil 1950). In the beginning, CNP acquired 22 tankers for FRONAPE (Fig. 3.3).

Two other projects had heavy investments by the CNP: the Mataripe Refinery and the Cubatao Refinery, both providing full income, being managed as industrial entities (Vargas 1964). In the 50s, during the direction of the CNP, studies

[9]In 1949, was renamed to Study Center and Petroleum and National Economy Defense (CEPDEN).

Fig. 3.3 Poster of the III National Petroleum Defense Convention, promoted by the CEDPEN—1952. *Source* Cartaz da ... (1952)

were carried out for the construction of the Cubatao Fertilizer Factory, "which later would manufacture the first basic petrochemical product in the country, that is, anhydrous ammonia, using the residual gases from the Presidente Bernardes Refinery" (Seabra 1965: 119).

With regard to improvement and professionalism, in 1952, CNP adopted what we can consider as one of the main initiatives to change course for teaching and for petroleum research in Brazil, and that only added to the work of training Brazilian know-how: Technical Improvement Supervision Sector (SSAT), which aimed to generate professional and own manpower. The details of this process will be visible in the next part of this book.

Before CNP made such a decision, the initiatives to train labor in the country—a need we have tried to demonstrate in the previous pages—were concentrated on a secondary plan, including the application of funds and the problems that went through the decades. It was secondary in the sense of the number of demands and pressures that bordered the politics and economics of national and international oil.

At that time, three solutions were, naturally, to expand the professional staff of our country, according to Oliveira Júnior's classification, molded to the theme presented here: (a) hire foreign technicians; (b) send young people to study in foreign schools; and (c) to develop education in the own country (Oliveira Júnior 1959). All of these solutions were absorbed by CNP and, subsequently, by Petrobras, as demonstrated below.

Each point mentioned above would have its advantages and disadvantages. With respect to point (a), costs were high (high salaries, language difficulties), and the techniques employed by professionals would differ greatly from one country to another, even more so as to the terms used in geology. We also have the problem of bringing foreigners into the country. In this regard, Oliveira Júnior classifies the lack of scientists and engineers as a universal phenomenon, using the work.

L'angoissante pénurie d'ingénieurs et téchniciens, from 1956, to justify the France case: "The number of engineers and technicians trained each year is notoriously inadequate. The workforce remains practically the same in the past 35 years, while needs have doubled or tripled" (Oliveira Júnior 1959: 49). The advantage would be that, in the case of the hiring of foreign professionals, they could teach the techniques to the Brazilians to form their own know-how.

The point (b) demanded investment beyond the budget allocated to the CNP for this purpose. In addition, by participating of courses, specializations or even a full training abroad (with a minimum of six months in general), the country would be temporarily without this workforce, which was already restricted, and did not guarantee permanence of the professional in the CNP.

Regarding point (c), the costs would be far beyond the sum of points (a) and (b), beyond which educational, economic, and political changes would be needed within the country. Of all levels, this would be the most difficult to maintain and put into practice, including the lack of professionals to teach the subjects related to oil studies (including Geology, Paleontology, Engineering, among others).

UM GRANDE EMPREENDIMENTO ECONÔMICO LANÇADO NO BRASIL

Visando solucionar o problema do petróleo, o Chefe do Govêrno envia mensagem ao Congresso Nacional propondo a organização de uma emprêsa mista de capital público e privado para a industrialização dessa fonte de riqueza do País — A íntegra da mensagem presidencial

Fig. 3.4 Headline of the Brazilian Newspaper from 07 December 1951. *Source* Um Grande … (1951), p. 9 (Translate of the Fig. 3.8—A great economic entrepreneur released in Brazil aiming to solve the oil problem, the Government Chief sends a message to the National Congress proposing the organization of a public–private capital mixed company to industrialize this rich source of the Country—The complete Presidential message.)

The CNP absorbed these three points. In the early years, the body concentrated more on points (a) and (b) and, in 1952, began to invest in point (c), which is described as the first training attitude of its own workforce and incentives to develop and improve their technique. It was then that the creation of the SSAT, previously mentioned, began.

To conclude this topic, we cannot fail to use the above descriptions and advance the points that Petrobras has absorbed. It invested in all points, but in different scales of the CNP, associating them and understanding their impacts in the structures of professional formation of the country, mainly by point (c). It has thus contributed massively to sectors and transformations in teaching in Brazil, mainly in the area of geosciences, in the construction and development of the technique, in the maintenance of its own equipment and in the formation of its know-how.

3.2 Petrobras and the Participation of Foreigners (1953–1961)

On January 31, 1951, Vargas became President again, this time by direct popular voting, and found a very different Brazil than he had previously governed. The class structure was more complex than the one existing in the New State, since it was mainly driven by the "twin processes of industrialization and urbanization," which "had increased and strengthened three sectors: the industrial, the urban working class and the urban middle class" (Skidmore 2010: 117). The goal of the new Vargas Government converged to the "accelerated industrialization as a condition of social progress […] and the state was armed with new institutions and instruments capable of making them viable" (Mendonça 1990: 333).

In the beginning of the 50s, with the developmentalist doctrine, in Brazil raises the necessary technologies for the achievement of the energy and transport goals, which complete the framework of Brazilian civil technology reaching international notability levels (Vargas 1994: 24).

Thus, the Vargas government (1951–1954), internally, "emphasized the need for state-owned enterprises as the basic instrument of an investment policy" (Skidmore 2010: 132). In December 1951, the government took to the national Congress the bill for the creation of a mixed (public–private) oil company, which would hold the "monopoly on oil exploration and all new refineries"[10] (Skidmore 2010: 132) (Fig. 3.4).

The debate over oil policy, fueled by the Petroleum Campaign, dominated public attention for decades, intensifying in 1945. The theme prevailed at the center of public discussions, more than any other subject (Skidmore 2010: 133), contributing to the creation of this state monopoly enterprise.

[10]"Although existing refineries were allowed to remain in the hands of private companies, and the distribution of petroleum products was left to the private sector" (Skidmore 2010: 132).

On October 3, 1953, through the Law No. 2,004, the company Petróleo Brasileira S. A. was created (Acronym or abbreviation: Petrobras), a mixed-economy corporation that had the objective the research, mining, refining, trading and transportation of oil from wells or shale from its derivatives, as well as from any related or similar activities (Brasil 1953). With this same law, the petroleum was monopolized[11] as a good of the Union "through the National Petroleum Council, as a guiding and inspection body," and "through the joint-stock company of Petrobras and its subsidiaries, incorporated in the form of the present law, as executing agencies" (Brasil 1953).

For Luciano Martins, this moment is considered a fertile field for the analysis of the dynamics of the actors and their ideologies during a phase of complex political reorientation of the society, since, for him, the nationalization of the petroleum It cannot be confused with the rise of the nationalism, but with its decline, from the second half of 50s, the nationalism became more a reactionary phenomenon, as a project and an instrument of mobilization, i.e., through it, the control of the nation against imperialism was sought. This role played by nationalism during this period facilitated the implementation of a developmentalism project (Martins 1976). The political and economic realization of this project occurred under the government of Juscelino Kubitschek (1956–1961), in which there was remarkable economic growth, especially after the expansion of industrial production.

It is also worth noting that Petrobras creation based on state investment (even in the case of the mixed economy, since this is the majority control of the Union) is not a pioneer in Brazil, and a few other examples can be mentioned: National Steel Company (1941), the Vale do Rio Doce Company (1942), the Álcalis National Company (1943) and the São Francisco Hydroelectric Company (1945).

At an accelerated pace, with a budget higher than that the one from CNP, Petrobras faced "a problem of considerable dimensions—the lack of specialized personnel to operate the entire industrial complex" (Fortes 2003: 2), such problem that, for many years, also made difficult the CNP work.

On April 2, 1954, by means of Decree No. 35.308, the Brazilian Constitution of the Petróleo Brasileiro S. A. was approved, having in the description of its art. 45 "that the company will contribute to the preparation of technical personnel and skilled workers, through specialization courses, granting aid, scholarships or other appropriate means" (Legislação brasileira …, s.d.: 255).

Petrobras began its activities in 1954, receiving the CNP collection and gradually absorbing it into its structure. With the CNP, the data up to 1952 indicated a total of 311 drilled wells since the beginning of its activities; "of this total, 180 produce oil, 24 gas and 107 are dry; 295 are located in the State of Bahia, 7 in Alagoas State, 4 in Sergipe State, 2 in the Territory of Acre State, 2 in Pará State and 1 in Maranhão State" (Vargas 1964: 134). It should be noted that the amount of oil produced was still considered relatively low to raise Brazil to the level of self-sufficient country in oil production.

Petrobras invested, then, in the expansion of the collection received. Part of this expansion was intended for the exploration and "short and

[11]"Such monopolization could only result from invocation of art. 146 [of the Federal Constitution of 1946], which says: 'The Union may, by special law, intervene in the economic domain and monopolize a given industry or activity. The intervention will be based on public interest and by limiting the fundamental rights guaranteed in this Constitution'. There may be *monopolization* by the Union of the ownership of the land *before* any authorization. There may be *monopolization*, by the Union, of land ownership, *after* and *independent* of the right of exploration. There may be *monopolization*, by the Union, of both rights or only the right of exploration, to be exercised or already with the industries installed, or installed and in progress. Law No. 2,004, dated 3 October, 1953, enumerated the attributions of the National Petroleum Council and created Petrobrás. Such a law could only be made with invocation, explicit or implicit, of art. 146 of the Constitution of 1946, since, in art. 1, established: 'They constitute a monopoly of the Union: I. The research and development of oil deposits and other fluid hydrocarbons and rare gases, existing in the national territory. II. The refining of the national or foreign oil.' […] law no. 2,004 created the monopoly (article 1) and could, before the Constitution of 1946, create it" (Miranda 1960: 523).

long-term preparation of specialized personnel from various professional levels and only constituting an agency with such specific powers, PETROBRÁS could fill the gap of school system in force in the country" (Oliveira 1961: 141).

> Within the "financial and human resources of the Nation", entrusted to it by law, PETROBRÁS did not have the immense capital, know-how and technology summed up and multiplied by the great international oil companies in the course of almost 100 years, around the world. I have rapidly expanded its industrial park to minimize or halt the heavy import of derivatives and finance exploration; promoted the specialization of dozens of technicians abroad; right here at the Center for Improvement and Petroleum Research (CENAP), [...] has acquired a large contingent of labor, not formed in Brazil by the curricula. And it threw itself into all our sedimentary basins in search of oil (Moura and Carneiro 1976: 307).

The idea that came from the CNP was to form "teams of Brazilian technicians that would comply with the growing needs" and "to reduce as much as possible the hiring of foreign professionals by gradually replacing them by national technicians" (Fortes 2003: 3). This replacement, indeed, did not happen immediately, as the growing internal demands and the expansion of the petroleum industry that arose at Petrobras required many professionals and the development of their own know-how, which could only be achieved with the contribution of foreigners at that moment. It was also intended to replace the foreigners with the Brazilians in the highest positions of Petrobras, thereby terminating the payments made, mostly in the US dollar.

In the internal organization of Petrobras 1955, the company operations plan included thirteen operating units in different areas (as shown in Fig. 3.5, p. 116), one of which was the Exploration Department (DEPEX), the unit of analysis in this book for two reasons: (1) the maximum superintendent position as Chief of the DEPEX belonged primarily to a foreign geologist; and (2) the department was responsible for the discovery of oil wells in the country and for their management. It was also in 1955 that Petrobras created its own agency of improvement and professionalization, named CENAP, as pointed out by the reference from Moura and Carneiro. Such an agency will be the study object of the third part of this book. For now, we will talk about the DEPEX.

In 1954, Petrobras took an attitude that would later change the history of oil in Brazil, as well as its advances. The prominent petroleum geologist Arville Irving Levorsen contacted the North American geologist Walter Karl Link (1902–1982), in order to hire him for the highest position of the Exploration Department, with the challenge of "organizing an Exploration Department, based on the most successful international companies" and informing Petrobras about "Brazil petroleum prospects" (Link 1961: 1).

In that period, Walter Karl Link was considered by his peers one of the six best geologists when it came to oil exploration (Eugênio Gudin … 1954). Link has a degree in Geology from the University of Wisconsin-Madison, United States, and gained extensive experience when working for the company *Standard Oil Co.* Louisiana, in Shreveport, and for *Standard Oil Co.* from New Jersey (Humprey and Sanford 1983: 1040) mapping oil deposits in Latin American countries such as Venezuela, Colombia, and Ecuador, among others (Dott 2011). At the end of June and beginning of July 1954, Link made a preliminary visit to Brazil, having a direct contact with Petrobras president, Colonel Juracy Montenegro Magalhães (1905–2001),[12] and two directors of the Company—Irnack Carvalho do Amaral (1905–1983)[13] and João Neiva de Figueiredo.[14] Local newspapers reacted in different ways to Link's hiring: on the one hand, they supported the presence of a recognized specialist to find oil in Brazil; on the other hand, the fact that the geologist was a former Standard Oil official contributed to the nationalists' view of him as a man with international rather than national interests. Despite the conflict of ideas

[12]Brazilian Military and Politician. He was the first President of Petrobras (from 2 April 1954 to 2 September 1954).

[13]Mining and Civil Engineer. He was the tenth President of Petrobras (from 30 June 1966 to 27 March 1967).

[14]Mining and Civil Engineer.

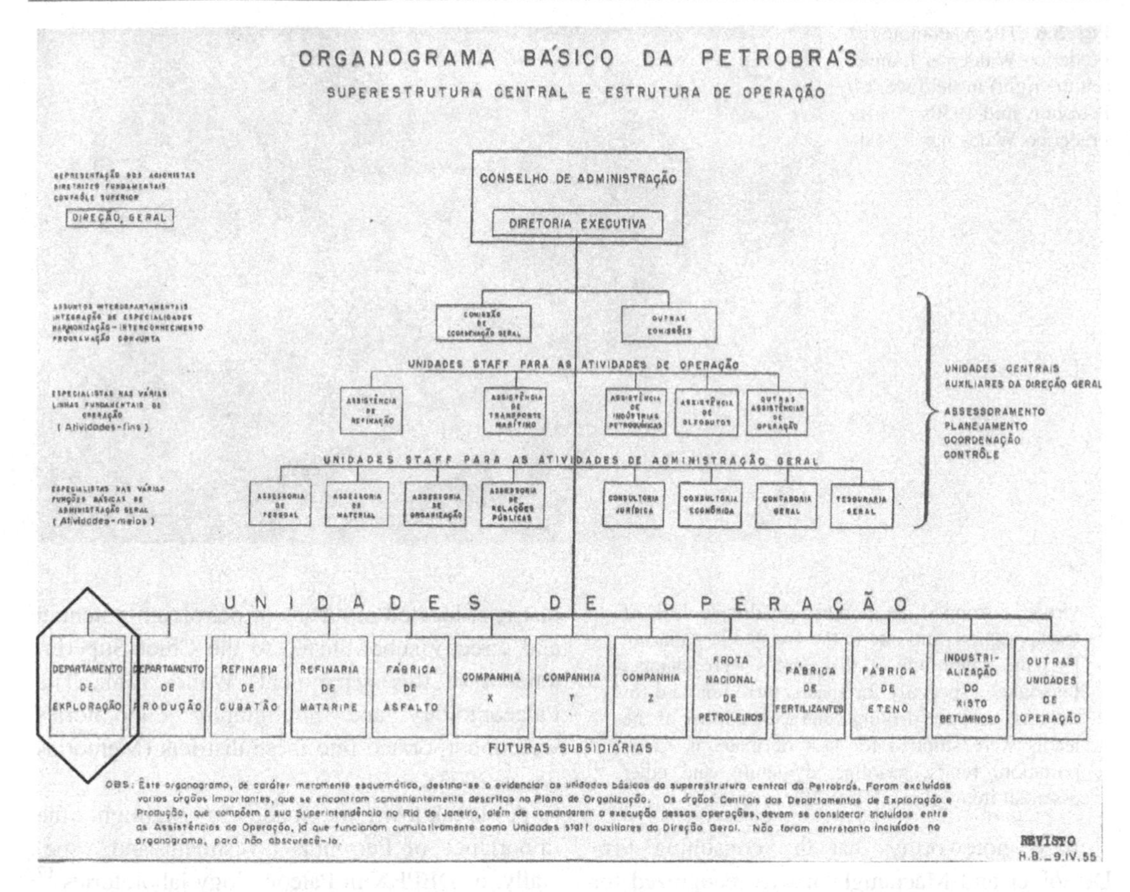

Fig. 3.5 Basil organizational chart of Petrobrás—1955. *Source* Organograma básico … (1955)

exposed in the media, for the first time a foreigner would assume a position of such importance in Brazil, as a Superintendent-Chief[15] of the Petrobras Exploration Department, with the responsibility of finding oil and seeking the so desired self-sufficiency.

Thus, Link hiring was also related to Petrobras desire to organize DEPEX "based on the most successful international companies, and to the fact that the Company wanted to know what oil possibilities in Brazil" (Link 1961: 1), previously unknown due to the extension of the territory and the geological knowledge of these areas. At that time, Petrobras prioritized exploration, since it was necessary to find oil to justify all the investments, as well as its own creation. In October 1954, when Link began his work at Petrobras, he described the situation he encountered:

> On the occasion of my arrival in Brazil, there was no exploratory group belonging to Petrobras in Rio office. CNP still controlled the holding through the consulting firm DeGolyer & Macnaughton, under contract with CNP. There were three surface geological classes in southern Brazil, one in Maranhão and two in the Recôncavo. There was a seismic team in the Recôncavo, two in Maranhão and two in Amazon. A pioneer well was waiting for material in Jacarezinho in the state of Parana, and two being drilling in Amazon. There was a gravimetric team in the Recôncavo, which was

[15]In order to understand the importance that we give here to the post of chief superintendent of a company, it is important to analyze the Brazilian social structure of the late fifties. For example, in the professional classification of the residents of the city of Sao Paulo, the positions of superintendent director, farmer, manager, lawyer, physician, priest and journalist are in the top rank of society, resulting from the progress of the processes of secularization and of commercialization of Brazilian society at the time. Mello, João Manuel Cardoso de Novais, Fernando Antonio. Late capitalism and modern sociability (Mello 1998).

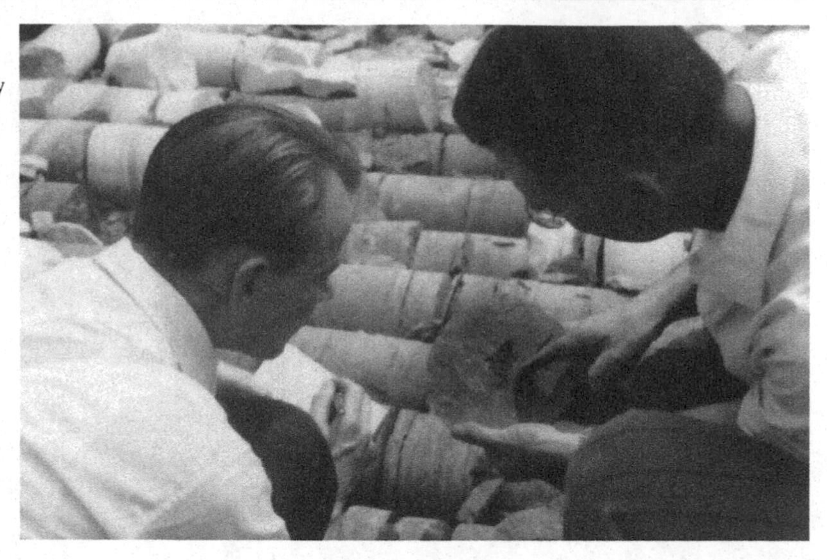

Fig. 3.6 The paleontologist Frederico Waldemar Lange (on the right) in field work by Petrobras mid-1950s. *Source* Frederico Waldemar … s.d

> virtually stopped 18 months ago due to lack of transportation, and one in the Tapajos in Amazon. […] the conditions of these areas were chaotic. Personnel, especially foreigners, were not paid for months, as were drilling contractors. Most of the teams were stopped for lack of material, transportation, funds, gasoline, dynamite, and other essential items (Link 1961: 2).

It is noteworthy that the consulting firm DeGolyer and Macnaughton was recognized for its work with geophysics and had as one of the main purposes to apply the recent technological advances of this area to find petroleum in Brazil. At first, it served as a service provider from 1944 until 1954, in Brazil, under a contract with CNP.

The conditions then found by Link were not at all favorable. It was necessary to organize the work carried out by CNP, now Petrobras, to reintroduce a new methodology, to offer better working conditions and to introduce new technologies. Parallel to Link's work, the DEPEX was organized in 1954 by the economist and administrator Hélio Marcos Pena Beltrão (1916–1997).[16] The plan was considered by Link as simple but concise, and therefore was maintained.

At the beginning of the DEPEX operation, its activities were divided and executed by regional districts, located in basins of petroleum potential and directly subordinated to the Chief Superintendent of this department, Walter Link. The Palaeontology and Stratigraphy Laboratories were incorporated into these districts (Memórias da … 2003).

A breakdown is made to highlight the importance of Petrobras investment and, especially, of DEPEX in Paleontology laboratories,[17] Stratigraphy and, right after Sedimentology. Major scientific advances and know-how formation were specifically related to Micropaleontology and Palynology carried out during this period by the company professionals and continued to expand, contributing significantly to new petroleum discoveries (Fig. 3.6).

[16]Graduated in the National Law School. He was Minister of Planning in the Military Dictatorship. He held the position of President of Petrobras from 19 March 1985 to 15 May 1986.

[17]For a long time, "researches in the area of Geology and Paleontology were restricted to field observations and point samples of sediments in the search for occasional fossiliferous findings. With the rapid expansion of exploratory research and the increasing volume of samples from drilled wells, there was a requirement for precise studies of the fragmented sedimentary material, aiming to describe the lithological characteristics and other characteristic elements. The microfossils found in the perforated sediments have been shown to be important as indicators of the sediment deposition environment and for its dating, reducing costs and optimizing oil exploration. From this recognition, Micropaleontology proved to be an important tool in the exploration of petroleum" (Memórias da… 2003: p. 8).

For José Luciano de Mattos Dias and Maria Ana Quaglino, Walter Link had as characteristics to institute a very ambitious exploration program (Dias and Quaglino 1993), but it was essential to change the course of exploration in Brazil. The American geologist complemented the Exploration Department with an organizational structure based on the American industry and collected previous studies and researches of the Brazilian sedimentary basins, also visited some of them for geological reconnaissance of the areas, deciding to focus their activities in two points: (a) in the development of the exploration in Bahia; and (b) drilling the largest Brazilian sedimentary basins (Amazon Basin, due to its vastness, and the Parana Basin, due to geographic and economic relevance). Both points indicated an "almost certainty of finding new oil fields" (Moura and Carneiro 1976: 308). Link also invested in the hiring of foreign geologists and geophysicists, in addition to obtaining internships for Brazilians in different universities abroad, in common agreement with CENAP.

For a better elucidation of the studies collected by Walter Link and the maps elaborated in the period, which showed the basins with greater chance of finding oil, we bring the map on the classification of the petroleum conditions of Brazil, elaborated in 1948, by Juarez Távora in the book Petroleum do Brazil, from 1955. It can be compared to the other maps still in this part of this book. In an attempt to confirm these possibilities, field teams were set up within the DEPEX and their respective districts to study and research of these basins.

It is reiterated that, from the beginning, both by CNP and Petrobras, the objective was to achieve self-sufficiency in oil production. This has become a challenge as no commercially exploitable wells were found and the country was under industrialization process, increasing the need for oil.

In 1958, Link insisted that one of Petrobras weaknesses in attempting to solve the problem of finding oil in the country was still the training of Brazilians in the formation of national technique, he had to recruit qualified personnel because of shortage (Passarinho 1958).

These statements are based on Link's report that, at the beginning of the DEPEX activities, the training of 50 Brazilians was conducted, and at the end of the study, most professionals abandoned their careers on the grounds that the work and the activities in the geology and geophysics area were too heavy for them, that they faced family problems, and that wages were low. On the other hand, Link emphasized that CENAP efforts were contributing to positive changes in this scenario (Link 1961: 2).

Even in the face of such difficulties, several inquiries were made by Link and his team. In 1959, during the 5th World Petroleum Congress, he presented the first results, pessimistic, about finding oil in the Brazilian territory (Smith 1978).

In view of these statements by Link, Petrobras began, in early 1959, to take steps to modify the critical situation that apparently starts in the country in the political, economic and social fields. It is then an investment in the technical–scientific exchange and the hiring of professionals, seeking a new opinion of the Brazilian sedimentary basins or new techniques to find oil in the territory. As an example, the arrival in Brazil of the geologist Claude de Lapparent (1920–1985),[18] Director of the Geology Department of de French Oil Company (Algeria) that visited the works developed by Petrobras, mainly in the Recôncavo Baiano region and in some technical agencies, as well as lectures, including at the Technical School of the Brazilian Army and the Engineering Club, both in Rio de Janeiro and related to farms in the Sahara (Famoso geólogo … 1959).

Subsequently, some foreign companies, such as those mentioned, that participated in this scientific exchange were dissatisfied with the lack of continuation of the work in Brazil, as their participation became increasingly limited. Petrobras was looking for another short visit to areas that are not very well known, with a greater

[18]Graduated from the Sorbonne in Sciences and the French Institute of Petroleum in Geology. Son and grandson of geologists, it is worth mentioning that the Treaty of Mineralogy written by his grandfather is a classic work in the field of geology.

emphasis on the study of existing work than on investigations to be carried out by foreign companies, i.e. the arrival of foreigners in the country would have restricted their orientation and consultation more than the proper execution of the exploratory services of other countries. Petrobras sought to form its own technique based on foreign information. Foreign firms were, therefore, dissatisfied with the company attitudes (Lange 1961).

It was in 1960, based on the results of the research conducted by Link in the Amazon Basin, that the situation was critical. The Amazon Basin was Petrobras and Link's main focus since the beginning of the activities to find new oil wells, with investments accounting for 60% of the total exploration budget (Link 1960). The site had also been indicated, as mentioned in the first part of this book, by the Israeli geologist Charles White, who concluded in the White report (1908) "that the intrusions of eruptive rocks into sedimentary formations make it unlikely that large petroleum, except in the Devonian lands of the Amazon Basin" (Pedreira 1927: 14, emphasis added). The Amazon Basin, on the map depicted earlier (Fig. 3.7), was again considered to be more likely to encounter oil (Fig. 3.8).

In the same year, compiling all data obtained in the research developed in the Amazon basin since 1955, and considering the geological factors as interpreted in the period, Walter Link concluded that after years of work of seismic, gravimetric, geological field, drilling, and also:

> […] aerial photograph; and aero magnetometry research in 60,000 km^2 in the Nova Olinda area; the drilling of 100 dry wells, and finally an investment of nearly 200 million dollars, it is impossible for me to reach another conclusion that large-scale exploration should either be suspended or drastically reduced to an experimental basis. Maybe ten years from now new geophysical techniques may have arisen that can solve the geophysical problem. The geology, however, will not change and I do not think the Amazon basin has the necessary geological requirements to become a major oil-producing region (Link 1961: 2).

On this path, Link made an estimate to find oil in the Brazilian sedimentary basins of a well with a possibility of production for each 275 drilled ones, adding that the technique and geological and geophysical knowledge were currently available (Link 1961). Finally, the geologist recommended that Petrobras should invest in other areas for exploration, but what would they be?

In August 1960, Link presented to the Petrobras' president, General Idálio Sardenberg, "a detailed report of the six years of research in Brazil […] [that] expressed the opinions of fourteen prominent geologists (six Brazilians and eight foreigners)" (Smith 1978: 137) belonging to DEPEX, the information described in the report was not encouraging. Link emphasized that research should be directed toward the continental shelf, as it would be Brazil new source of oil and natural gas, or investment in other countries. The Link Report,[19] considered confidential, leaked to the press and had national and international repercussion. "The report fell as a bomb in public opinion, because it questioned the existence of large oil deposits in Brazil, one of the beliefs of the nationalist sectors" (Costa, s.d.).

> The "confidential" report appeared in several newspapers in Rio de Janeiro in mid-November 1960, and Gabriel Passos[20] made his point in the Chamber of Deputies, accusing Link of having "sabotaged" the exploratory effort of Petrobrás because he was secretly still in the service of Standard Oil of New Jersey. Passos alleged that Link always ordered that "the drilling was interrupted as soon as the oil evidence was observed". He even said that Link had been hired by mistake and that Petrobrás had hastened to clarify Link's situation in the company and ended by saying: "Petrobras opinion is that oil in Brazil is economically viable" (Smith 1978: 137–138).

[19]According to Azevedo and Terra (2008: 381), the report is composed of three letters and "in popular parlance, it became known as the 'Link Report'". They are followed by a fourth document containing the proposed exploratory budget for 1961.

[20]Gabriel de Resende Passos (1901–1962) born in Itapecerica (Minas Gerais State). In 1924, he had a bachelor degree in legal and social sciences from the Belo Horizonte Law School. He acted as a lawyer, journalist and politician, having been elected a federal deputy. He also served as General Attorney of the Republic (1936) and Minister of Mines and Energy (1961).

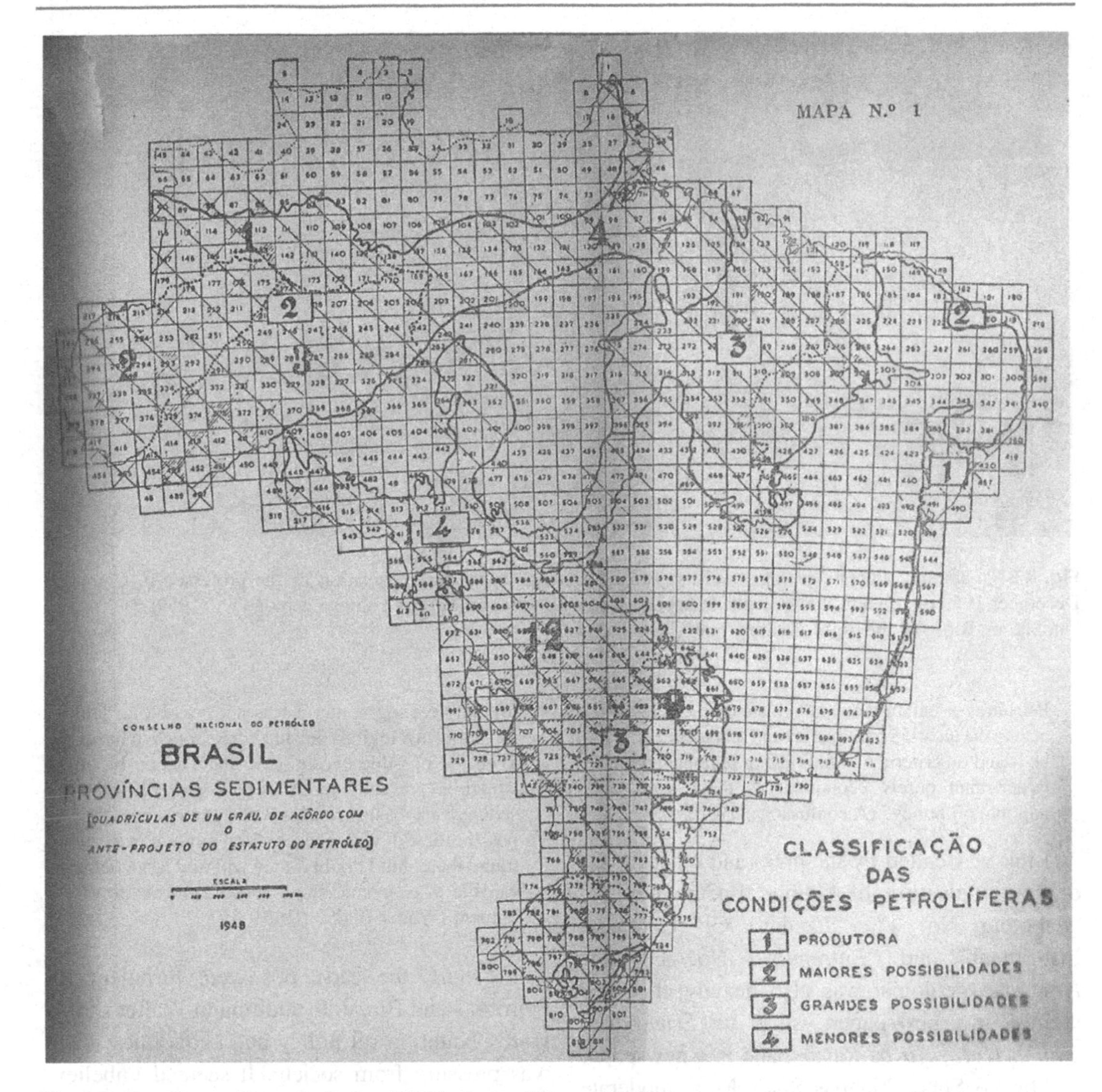

Fig. 3.7 Brazilian sedimentary basins (1948). *Source* Távora (1955)

Much of the information, such as those described above, has become more of an attempt to appease the negative news about Brazilian oil and find a culprit—in this case, Walter Link—than to take all data into account and invest in new solutions to the problem. Still, in this case, the discussion is not restricted to Petrobras alone and to the problems of internal management of the company, mainly in relation to the opinion of public persons, but becomes a national question. As Walter Link put it, "political speeches, articles or writers' books, attacks on men seeking to unravel the geology and problems of their land, will not be able to change existing geological conditions or put oil underground" (Walter Link … 1961, p. 17). A specific case of public criticism is that of Deputy Ferro Costa, who goes beyond the considerations in the Report, not accusing Link, but focusing on Petrobras internal irregularities:

> Wouldn't be Mr. Link, by this point of the events, being used as scapegoat? The campaign against Mr. Link will not be a smoke screen to cover irregularities of other sector from Petrobras […] Why has Petrobras not yet developed a national refining plan? And the question of the production of basic lubricants by the Mataripe Refinery? And the financial situation of the National Tanker Fleet? will

Fig. 3.8 "Famous French geologist visits Brazil"—December 1959 on the left, the geologist Lapparent when landing in Rio; on the right, visiting the CENAP, he listens to the explanation of the professor F. Campbell Williams. *Source* Famoso geólogo … (1959: 5)

> Petrobras orientation not only relate to research, but also to the location of refineries and oil pipelines, not be based on criteria of a political or regional nature rather than purely economic? Is there really a national oil policy? (A confusão … 1960).

Link was called by the press and other means of communication of "Public Enemy No. 1", "Saboteur No. 1", "Standard Oil Company Instrument" and "*Entreguista* No. 1". The American geologist was also heavily criticized mainly in "newspapers such as *Semanário, Última Hora, Novos Rumos*, and in a newspaper in Maceió State, Alagoas State. More moderate attacks occurred in Paraná State and São Luiz State, Maranhão State" (Link 1961: 2). In his defense, he categorized such attitudes as continuous personal attacks. The magazine *O Cruzeiro* defined well the situation of Link: "Fought by some, applauded by others, the American geologist left Brazil at the end of his contract with Petrobrás" (Walter Link … 1961: 17), in December 1960.[21] In 1961, before returning to United States, Walter Link described that:

> my only feeling is that it was not possible to find oil in the vast regions outside the Recôncavo basin. With the possible exception of the Tucano basin, which is presently under exploration, the other sedimentary basins remain with little encouraging possibilities. I will not say that any oil will be found there, but I doubt if they will find large fields capable of economic and commercial scale development (Walter Link … 1961: 17).

In short, the early 60s were turbulent for Petrobras and Brazil. In addition to Walter Link's words haunting oil policy and economics, there was pressure from society. It seemed unbelievable that a country with such a territorial extension did not have marketable oil. To aggravate the situation, Brazil's oil exploration data, compared to other countries with smaller territorial extensions, were discouraging. Statements by President Juscelino Kubitschek, a few months before the Link Report was published, asserted that Brazil had exceeded the oil production target, put the government plan in disrepute in that regard (Atingidas e … 1960).

The representation would come by data and map (Fig. 3.9, Table 3.1).

Due to the repercussion of the Link Report, and in view of the contacts that Petrobras made with other international industries in order to

[21]A few months earlier, Petrpbras attempedattempted to renew Link's contract for another two years, but he opted to close his deal with the company because of the results described and growing personal attacks.

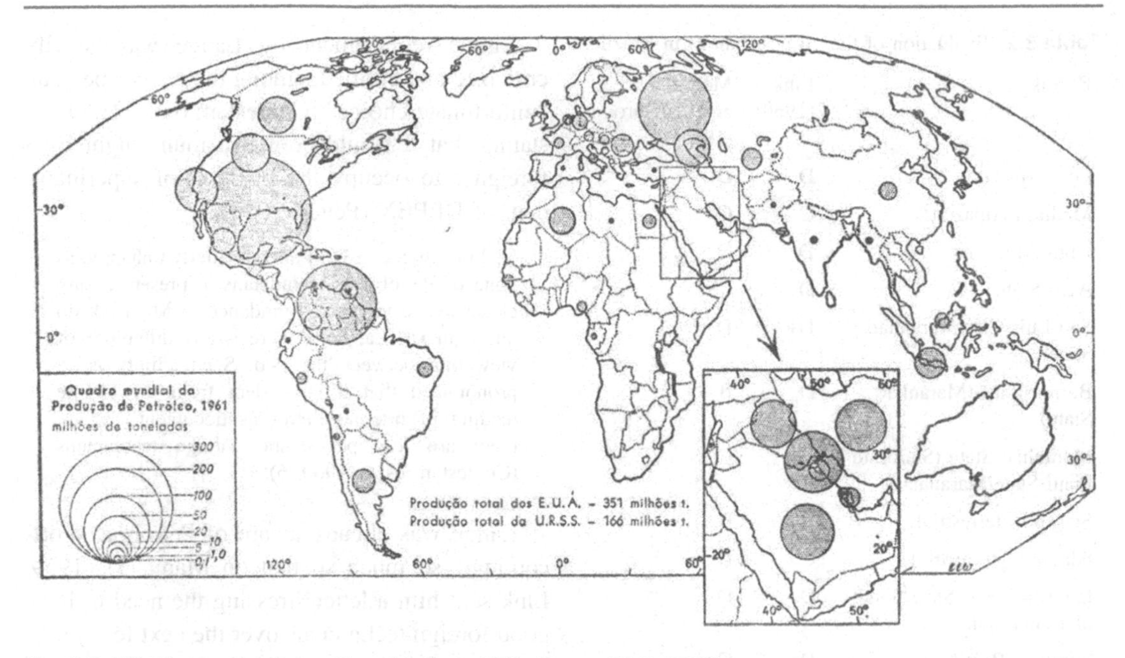

Fig. 3.9 World scenario of the oil production—1961. *Source* Odell (1966)

Table 3.1 World production of crude oil—1955/1960

Countries	Mean daily production (thousands of barrels)					
	1955	1956	1957	1958	1959	1960
United States	6,807	7,151	7,170	6,709	7,042	7,077
Mexico	250	257	253	276	290	264
Venezuela	2,157	2,457	2,779	2,606	2,768	2,905
Brazil	6	11	28	52	65	71
Holland	20	21	29	31	34	35
France	17	25	28	28	31	39
Germany	61	68	77	86	99	108

Source Produção mundial … (1961: 1)

reverse the negative post-report situation, in December 1960, French technicians were hired for "a new geological survey of the regions of the country where the existence of oil is admitted since one does not have the elements to challenge the Link Report and one does not intend to accept the conclusions of this report as definitive" (Petrobras vai … 1961, p. 3). In March 1961, led by Petrobras, the geologists Pedro de Moura and Décio Savério Oddone reevaluated the Link Report (Dias and Quaglino 1993: 119) that established the classification of the basins in:

He classified with "A" the basins that had commercial production, in which the exploratory efforts had to continue. They were the basins where all the mentioned conditions were present. As "B" were basins that had the existence of matrix rocks, but where there could be no porous or fractured rocks or no occurrence of structures or other favorable geological conditions. In those basins, the geological factors indicated that there were possibilities of commercial discovery. These were areas where exploration should be continued. As "C" the marginal basins, which would have weak or limited characteristics of matrix rocks and little evidence of the existence of reservoirs or structures, in which it had made a great exploratory

Table 3.2 Evaluation of the oil possibilities in Brazil

Basins	Link (1960)	Moura and Carneiro (1976)
Low Amazon	D	D
Medium Amazon	C	C+
High Amazonas	D	D
Acre State	D+	C−
São Luís City/Maranhão State	D+	D
Barreirinhas (Maranhão State)	D	B
Maranhão State (Sudoeste Piauí State/Maranhão)	D−	C
Sergipe—terrestrial	C−	B
Alagoas—terrestrial	C−	B
Espírito Santo State/South of Bahia State	D	D
South of Brasil	D	C−
Recôncavo Basin	A	A

Source Moura and Carneiro (1976: 331)

effort, without results. In these regions, some exploration could still be carried out.
As "D" the basins in which there was no possibility of oil existence, because they did not present the existence of matrix rocks. In these basins, no further exploration work should be carried out (Costa, s.d.: 19).

Some letters were accompanied by a "+" or "−" symbol according to the diagnosed results, not influencing much in the final results. Thus, geologists have come to the following conclusion regarding the Link Report (Table 3.2).

By using this table, it is possible to verify that data differs little, with the specific exception of some basins. Thus, Petrobras was reluctant to accept Link's words, refusing to invest precisely in research on the continental shelf and insisting on the exploration of the territory even with the presentation of the above data. Only years later Petrobras recognized that Walter Link was right.

On January 1, 1961, at Link's appointment, the paleontologist Frederico Waldemar Lange was appointed to the post of Chief Superintendent of the DEPEX Exploration Department (formerly occupied by Link), which led to a new revolt in the press. Link's replacement by Lange was heavily criticized and entitled, among the newspapers, an "unfortunate choice" (Contestam os … 1960: 6), stating that it would be a Brazilian taught by a foreigner to occupy the position of superintendent of DEPEX (Peyerl 2010).

> […] the choice of Dr. Lange is utterly unhappy, as none of Petrobras's technicians at present is any closer to the views and guidance of Mr. Link do que than Dr. Lange, if there is any difference of viewpoint between the two. Such affinity is so pronounced that it is evident from the simple reading of internal Petrobrás documents, where there are less pessimistic foreign technicians (Contestam os … 1960: 6).

Lange was already aware of Petrobras shortcomings, so much so that on March 31, 1959, Link sent him a letter stressing the need to have good foreign technicians over the next few years, hoping that during that time the "best Brazilians" would develop enough to take on the jobs within Petrobras (Link 1959). Such proposal from Link was implemented by Petrobras through DEPEX, a decision that did not lack criticism.

In the period in which Lange was Chief Superintendent of DEPEX, he "performs administrative functions, proposes facilities of other laboratories in the districts, and also studies deeper in the region of Amazonas" (Peyerl 2010: 75), partially at the insistence of Petrobras. The contact with other countries contributes to technical visits and internships abroad, in locations such as Institut Français des Pétroles, Compagnie Génerale de Geophysique, Agip Mineraria/ENI (Italy), Amt fuer Bodenforsschung, Compagnie Française des Pétroles, among others.

From the Frederico Waldemar Lange collection, which has much information on the period in which he occupied the maximum position of oil exploration in 1961, it was possible to understand the participation of foreigners and the continuous construction of Brazilian know-how. Through a survey carried out in the letters and reports of such collection, especially in the early 60s, the main places of internship and visits of Brazilians were concentrated in the following countries (Organizational Chart 3.1).

Organizational Chart 3.1 Main countries visited and with internships by Brazilians (1958–1965). *Source* elaborated by the author

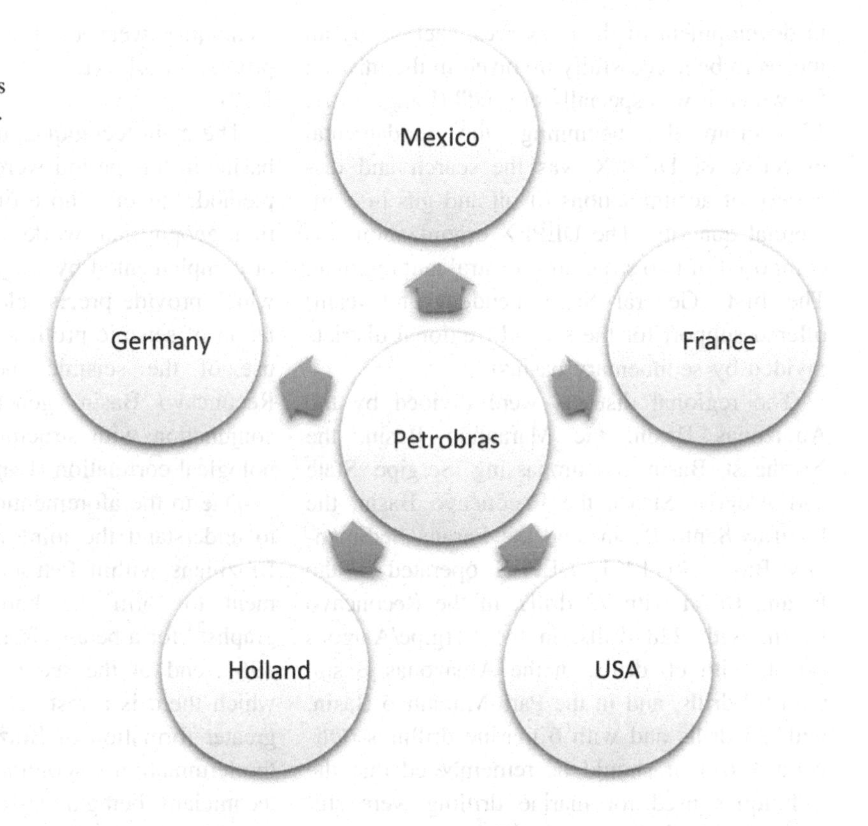

Petrobras then, through DEPEX, strengthened relations with these countries and with others in search of new technologies and improvement of their own know-how. The relationship that France has established with Brazil, especially in the contribution of the teaching structure in the Petroleum Area, especially for the development of CENAP and internships abroad.

In August 1961, part of the DEPEX members and other Petrobras professionals visited the *Ente Nazionale Idrocarburi* (Eni)[22] in Italy. The reason for the trip was linked to the analysis of general aspects of geological and geophysical exploration, marine drilling and eventual association for the joint exploration of petroleum abroad (Lange 1961). It was at this pace of visits and research that Petrobras strengthened its work and sought to build its know-how. In 1961, the Brazilian oil industry did not yet have a method capable of anticipating the possibility of developing sub-commercial wells or petroleum products (Amaral 1961). Therefore, it was necessary to seek technology abroad and adapt it geological characteristics of the country.

[22]The company was founded in 1953 by the Italian government with the aim of promoting and developing a national energy strategy.

3.3 The Oil Exploration Department (DEPEX) Seen by Numbers

In 1961, the book *Aspectos Econômicos de exploração de petróleo no Brasil* [Economic Aspects of the Exploration of Petroleum in Brazil], published by Frederico Waldemar Lange, is published. The invitation for the formulation of such work was made by the economist Hélio Beltrão (1916–1997), who had large participation, as already mentioned, in the organization of the DEPEX.

The fact that the petroleum does not contribute with 50% of the energy produced in the country was one of the reasons why Petrobras invested part of its budget in its exploration, "that is, in research for the discovery of petroleum and in development of their reserves, seeking by all means to be successfully involved in the mission for which it was specially created" (Lange 1961: 12). From the beginning, the fundamental objective of DEPEX was the search and discovery of accumulations of oil and gas in commercial quantity. The DEPEX organization was composed of two structures: central and regional. The first (General Superintendent and team) offered support for the second (regional districts divided by sedimentary basins).

The regional districts were divided by the Amazonas Basin, the Maranhão Basin, the Northeast Basin (encompassing Sergipe State and Alagoas State), the Recôncavo Basin, the Espírito Santo Basin, and the Paraná Sedimentary Basin. In 1961, DEPEX operated in the Paraná Basin with 72 drills, in the Reconcavo Basin, with 228 drills, in the Sergipe/Alagoas Basin, with 60 drills, in the Amazonas Basin, with 97 drills, and in the Para-Maranhão Basin, with 23 drills and with 6 marine drillings (Oliveira 1961). It should be remembered that the techniques used for marine drilling were still under development, and that Brazil was later an innovator, constructing new technologies for exploration on the continental shelf. For that reason, and for the analysis of documents in the Frederico Waldemar Lange collection, we are restricted here to the study of the sedimentary basins of the Amazonas and Maranhão, which corresponded to the same district, of the Northeast (Sergipe and Alagoas States), the Recôncavo and Paraná Basins, because, at the time studied, investments/researches were larger in those basins.

The study of each sedimentary basin presented different peculiarities and special exploration problems. For example, the Amazon Basin was difficult to penetrate, transportation was made by watercourses and there were geological problems of intrusive rocks and basalt flows, their thickness could reach many hundreds of meters (Lange 1961). In 1958, Walter Link pointed out that: "the problem of diabase in Amazonas and southern Brazil challenges the attempts to work structures with seismography; practically all the imaginable equipment and techniques were used in the solution of such problem [...]" (Lange 1961: 18) (Maps 3.11, 3.12).

The main techniques used in the sedimentary basins in this period were related to the seismic method, "mainly those of low frequency refraction, geophysical works that are associated with or complemented by deep stratigraphic drilling," which provide precise elements for the interpretation of seismic profiles (Lange 1961: 20). The use of the seismic method, mainly in the Recôncavo Basin, generated good results in conjunction with structural drills and by palynological correlation (Lange 1961).

Due to the aforementioned set of factors, and to understand the joint work of foreigners and Brazilians within Petrobras Exploration department to form the know-how, we produced graphs[23] for a better visualization of this work in 1961, end of the second part of this book, in which there is a restructuring of the DEPEX, a greater formation of Brazilian professionals and the termination of contracts with several foreign technicians, being a reassessment of the technical potential that the department had.

Graphs 3.1 and 3.4 represent the work within the DEPEX per analyzed Basin, specifically in areas of geology and paleontology. Graphs 3.5 and 3.8 represent well the tactic adopted by Petrobras to train Brazilian manpower, in order to replace the foreigners in the future. Graphs 3.9 and 3.10 show the same case, but in the paleontology area, Graphs 3.11 and 3.12 in the palynology area (we use the term used back then "palinologistas", replaced by "palinólogos"). Finally, we built Graph 3.13, which indicates the nationality of the foreigners who worked at Petrobras in 1961, and who may have been in Brazil for more than 2 years, thus demonstrating the variety of technical imports that we had for the formation of the petroleum know-how.

[23]Graphs 3.1, 3.2, 3.3, 3.4, 3.5, 3.6, 3.7, 3.8, 3.9, 3.10, 3.11, 3.12, and 3.13 were elaborated by the author based on data analyzed and compiled from the report entitled "Regional Organization and Situation of Technicians in the Exploration Department", signed by Frederico Waldemar Waldemar Lange in 1961.

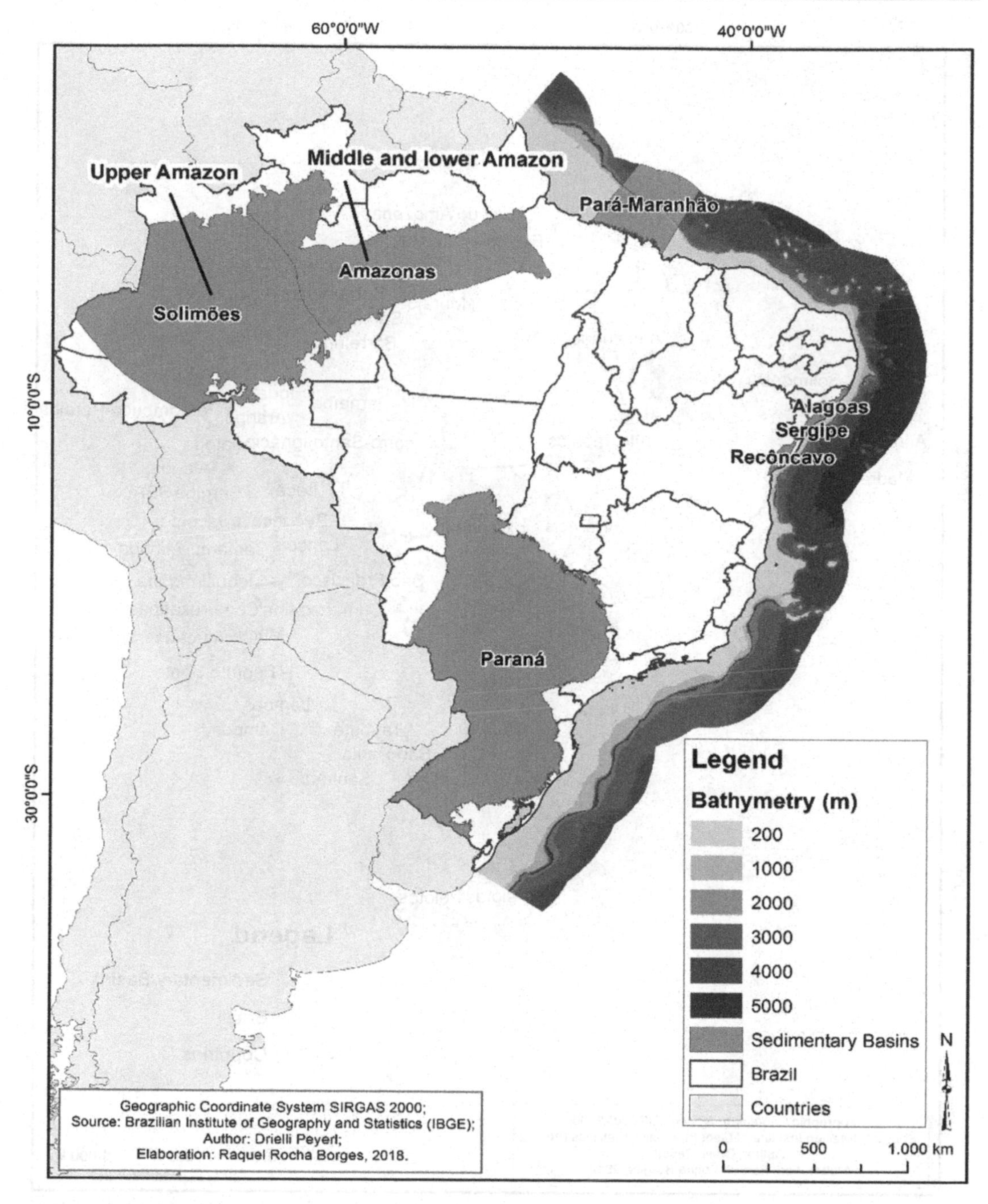

Map 3.1 Exploratory drillings made in the Brazilian sedimentary basins by the DEPEX—1961. *Source* elaborated by Raquel Rocha Borges (2018)

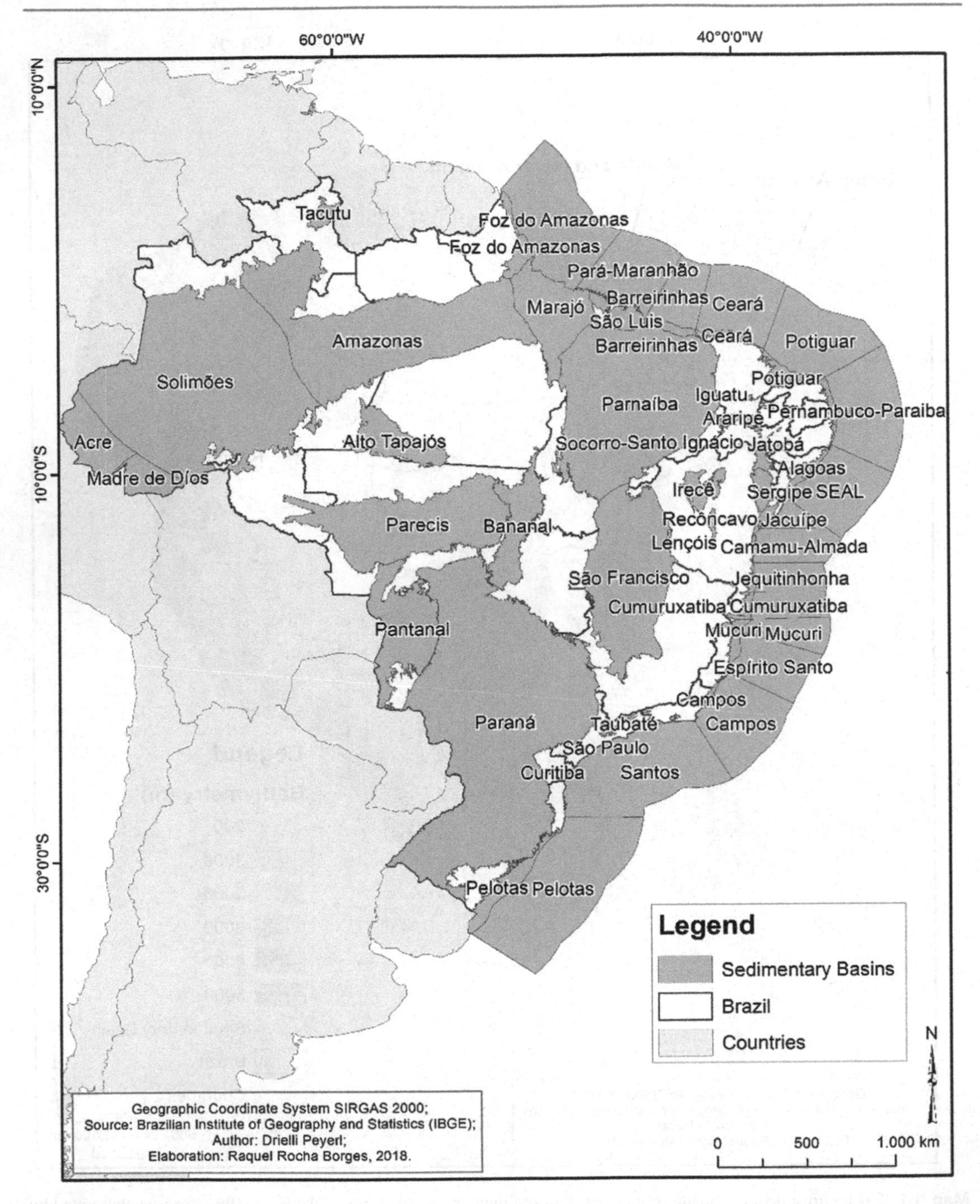

Map 3.2 Current Brazilian sedimentary basins. *Source* elaborated by Raquel Rocha Borges (2018)

Graph 3.1 Paraná Basin (geologists)—1961

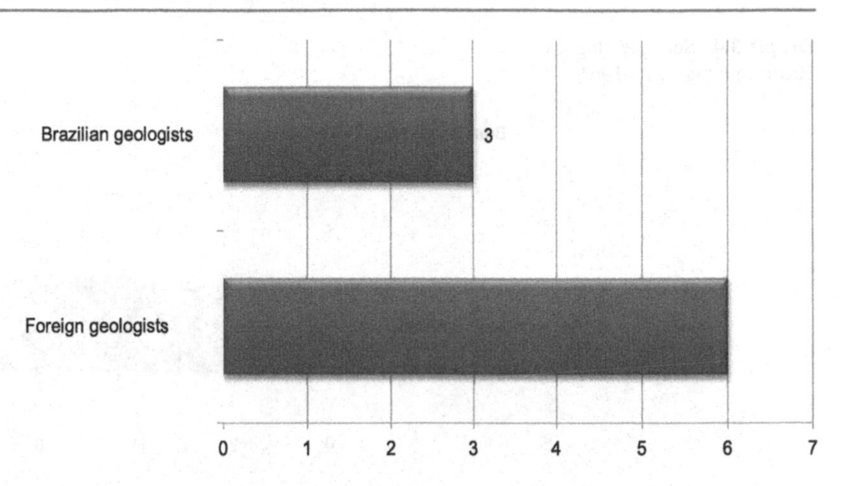

Graph 3.2 Amazonas/ Maranhão Basin (geologists)—1961

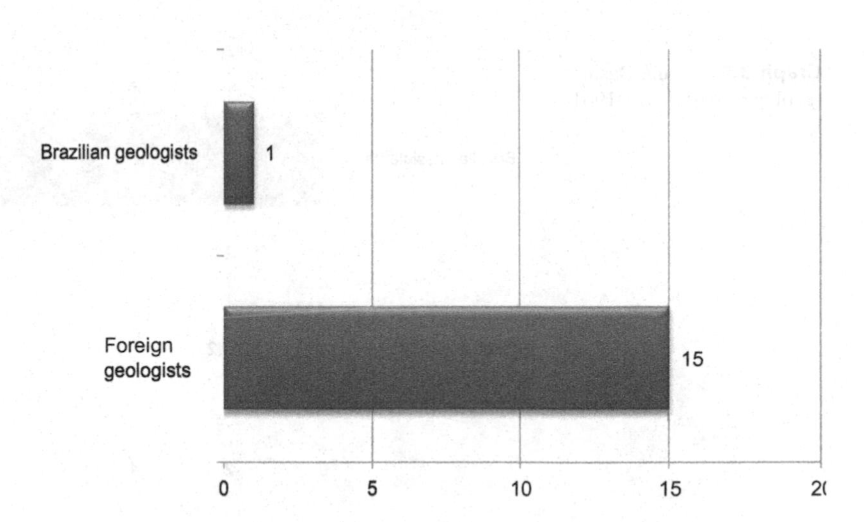

Graph 3.3 Recôncavo Basin (geologists)—1961

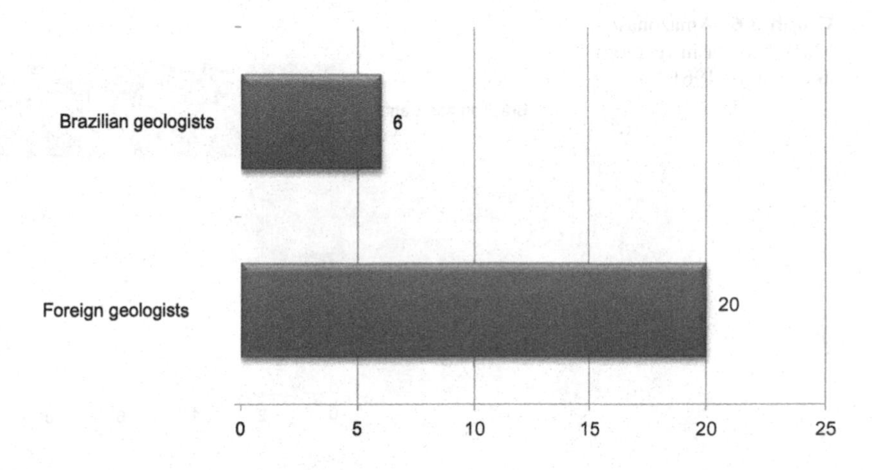

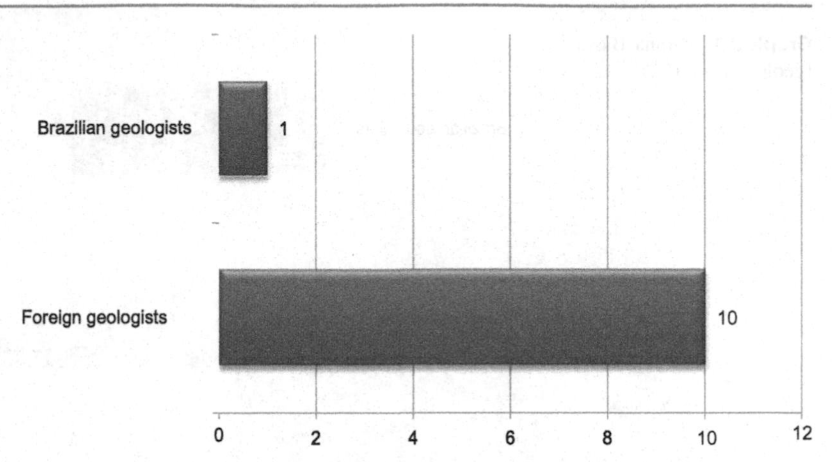

Graph 3.4 Sergipe/Alagoas Basin (geologists)—1961

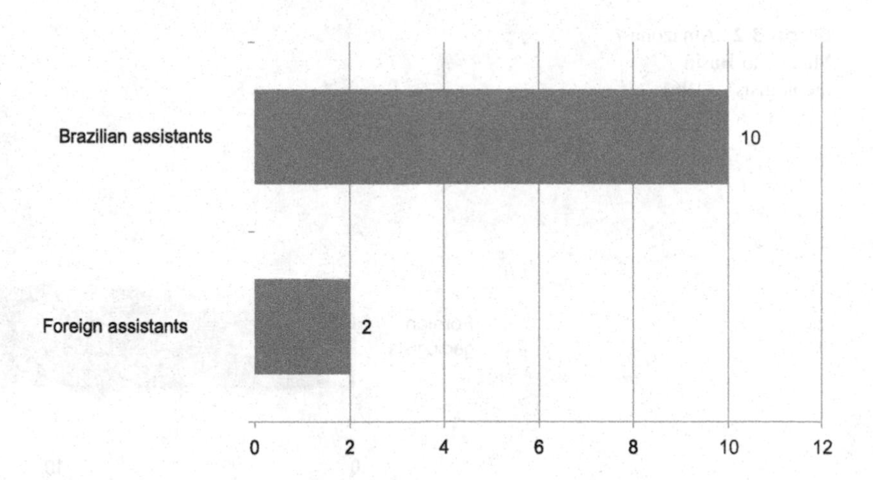

Graph 3.5 Paraná Basin (geology assistants)—1961

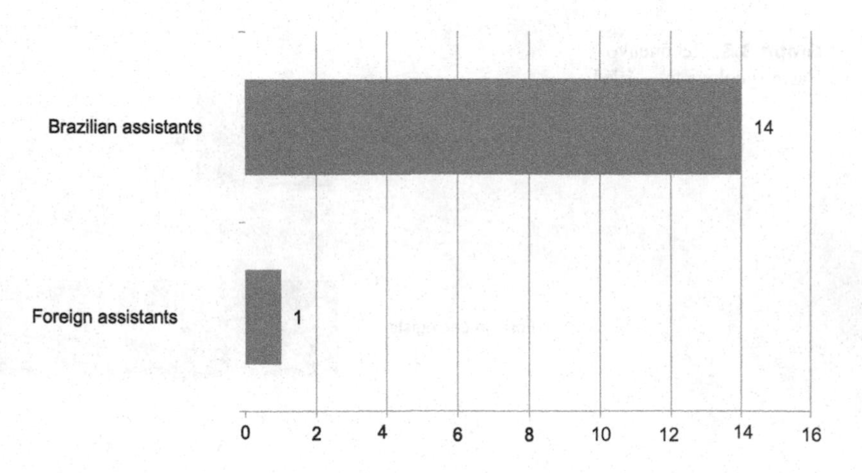

Graph 3.6 Amazonas/Maranhão Basin (geology assistants)—1961

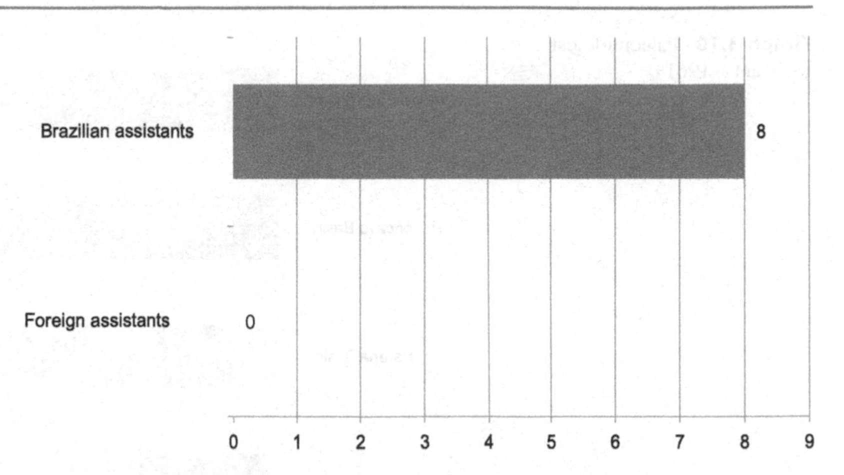

Graph 3.7 Recôncavo Basin (geology assistants)—1961

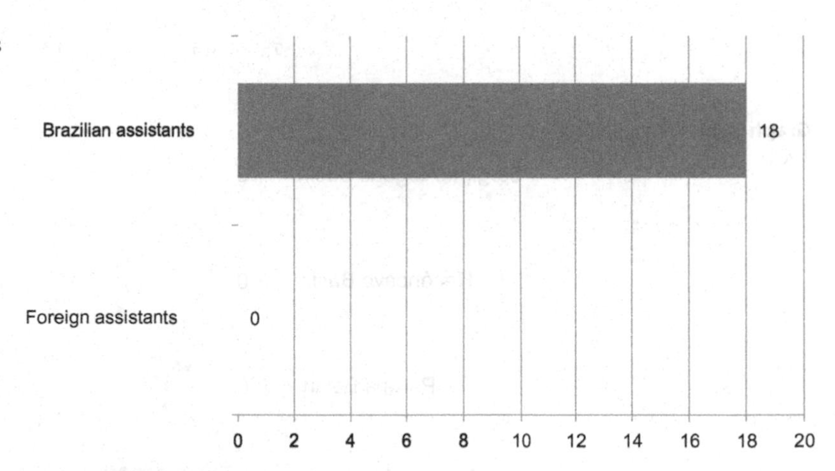

Graph 3.8 Sergipe/Alagoas Basin (geology assistants)—1961

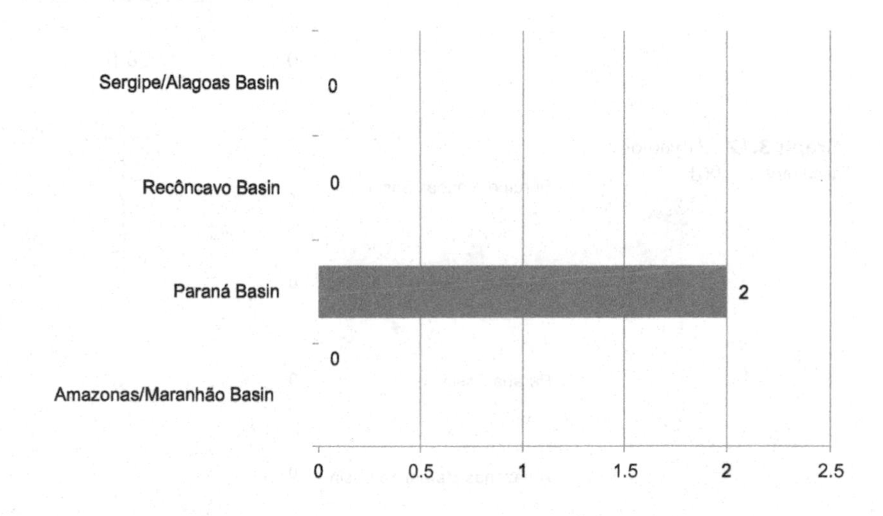

Graph 3.9 Brazilian paleontologists—1961

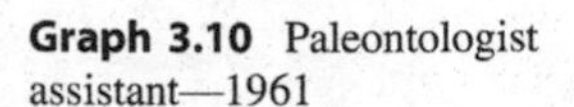
Graph 3.10 Paleontologist assistant—1961

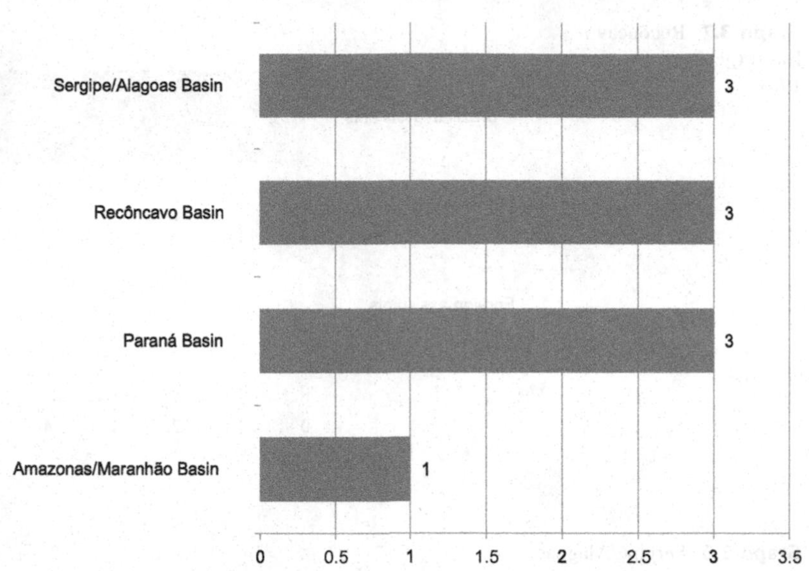

Graph 3.11 Palynologists—1961

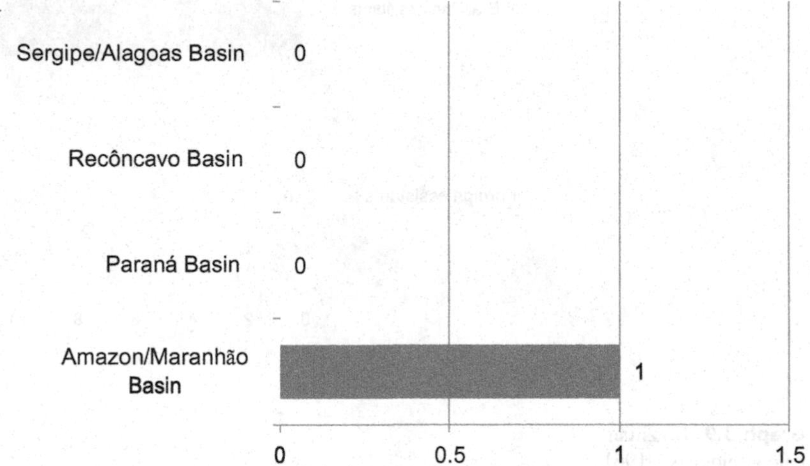

Graph 3.12 Palynology assistants—1961

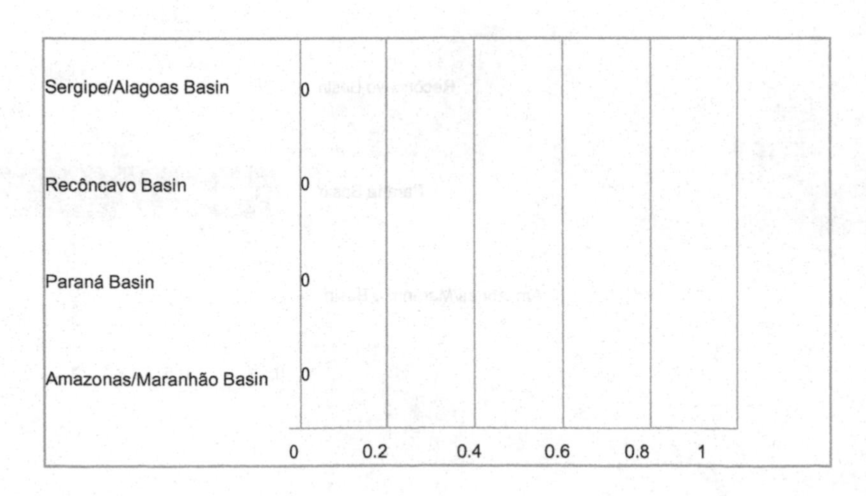

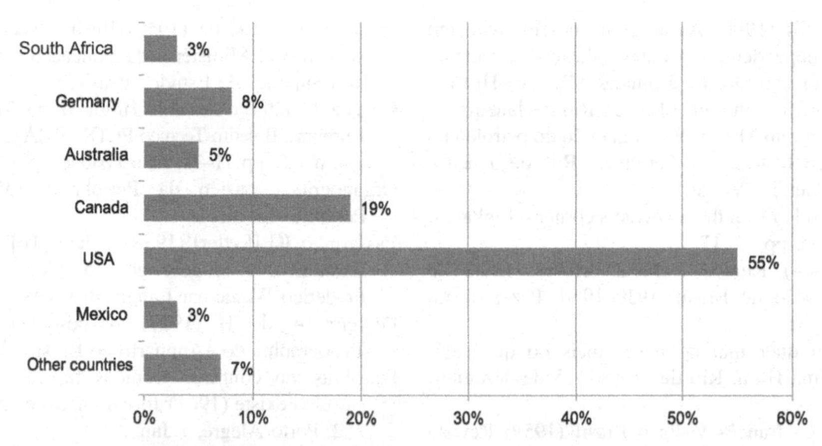

Graph 3.13 Formation countries of the foreign geologists (We emphasize that the majority of the geologists trained in the United States attended the following Universities, among others: University of California, University of Arkansas, University of Nebraska, Columbia University and University of Texas. The percentage of 7% is from countries such as England, Poland, France and others. These graphs demonstrate that the manpower of Petrobras began to be formed, for the most part, by Brazilians. We would like to point out that a greater understanding of how CNP and Petrobras managed to build their professional staff, a highlight item of this book, is in the third chapter. The graphs also show concrete data from Petrobras concerning the use of foreign labor for the learning of Brazilians, as there is a greater number of Brazilian geology assistants compared to foreign geologists working in the country.)

References

A confusão do petróleo (1960) Jornal do Brasil, 2 Dec

A questão do petróleo (1979) Política Mineral e Energética. Jornal do Geólogo, October/November/December

Abreu SF Petróleo (1948) Boletim Geográfico, ano 6, n 62, May

Andrade AMR (1999) Físicos, mésons e política: a dinâmica da ciência na sociedade. São Paulo: Hucitec; Rio de Janeiro: Museu de Astronomia e Ciências Afins

Atingidas e ultrapassadas as metas do atual governo (1960). Jornal O Jornal, 2 Feb

Azevedo RLM, Terra GJS (2008) A busca do petróleo, o papel da Petrobras e o Ensino da geologia no Brasil. Boletim de Geociências da Petrobras. Rio de Janeiro 16(2):373–410

Bastos PPZ (2004) O presidente desiludido: a campanha liberal e o pêndulo de política econômica no governo Dutra (1942–1948). História Econômica & História de Empresas 7:99–136

Bastos PPZ (2006) A dinâmica do nacionalismo varguista: o caso de empresas estatais e filiais estrangeiras no ramo de energia elétrica. In: Encontro Nacional de Economia. Anais... Salvador: Associação

Brasil. Decreto-Lei n° 538, de 07 de julho de 1938. http://legis.senado.leg.br/legislacao/listaPublicacoes.action?id=103275&tipodocumento=dEl&tipoTexto=PUB. Accessed Jan 2013

Brasil. Decreto-Lei n° 1.143, de 09 de março de 1939. http://legis.senado.leg.br/legislacao/listaPublicacoes.action?id=10299&tipodocumento=dEl&tipoTexto=PUB. Accessed Apr 2013

Brasil. Art. 153 da Constituição Federal de 1946. www.jusbrasil.com.br/topicos/10614743/artigo-153-consituicao-federal-de-18-setembro-de-1946. Accessed June 2013

Brasil. Decreto n° 29.006, de 20 de dezembro de 1950. http://legis.senado.leg.br/legislacao/listaPublicacoes.action?id=161095&tipodocumento=dEC&tipoTexto=PUB. Accessed May 2013

Brasil. Lei n° 2.004, de 03 de outubro de 1953. www.planalto.gov.br/ccivil_03/leis/L2004.htm. Accessed Jan 2013

Cartaz da III Convenção nacional de defesa do Petróleo, promovida pelo Centro de Estudos e defesa do Petróleo e da Economia Nacional (Cedpen) (1952). Rio de Janeiro: [s.n.]. http://memoria.petrobras.com.br/upload/depoentes/maria-augusta-de-toledo-tibiria-miranda/campanha-o-petroleo-enosso/5663/Hv032FT007_original.jpg. Accessed Jan 2013

Coelho WTS (2009) O monopólio estatal do petróleo no Brasil: a criação da Petrobrás. Revista História, imagem e narrativas, n 8, April

Cohn G (1968) Petróleo e Nacionalismo. Difusão Européia do livro, São Paulo

Contestam os autores que a Petrobrás se oriente por determinados técnicos (1960). Jornal O Globo, 1 Dec

de Mendonça SR (1990) As bases do desenvolvimento capitalista dependente: da industrialização restringida à internacionalização. In: Linhares MY (ed) História Geral do Brasil, 9th edn. Elsevier, Rio de Janeiro

Dias JLM, Quaglino MA (1993) A questão do petróleo no Brasil: uma história da Petrobrás. Rio de Janeiro: Fundação Getúlio Vargas

Dott Jr RH (2011) From the archivist's corner – Linkages. The Outcrop, pp 14–17

Draibe S (2004) Rumos e Metamorfoses: Estado e Industrialização no Brasil, 1930–1960. Paz e Terra, Rio de Janeiro

Eugênio gudin quer mandar muito mais do que café! Jornal Última Hora. Rio de Janeiro, 15 de dezembro de 1954

Famoso geólogo francês visita o Brasil (1959) Revista PETROBRÁS, Rio de Janeiro, ano 6, n 153, 1 Dec

Fortes AP (2003) CENAP - Petrobras: uma breve memória 1954–1964. Petrobras, CEnPEs/Petrobras Library

Humprey WE, Sanford RM (1983) Walter Karl Link (1902–1982). Bulletin of the American Association of Petroleum Geologists. Tulsa 67(6):1039–1040

Lange FW (1961) Aspectos Econômicos da Exploração do Petróleo no Brasil. Instituto Brasileiro de Petróleo, Rio de Janeiro

Link W [Letter] (1959) Rio de Janeiro [para] Lange, Frederico Waldemar. 1 f. Transferência. Archive Frederico Waldemar Lange, Box 114

Link W [Letter] (1960) Rio de Janeiro [To] Magalhães, Juracy. Bahia. 6 f. Archive Frederico Waldemar Lange, Box 108

Link W (1961) Exploração - PETROBRÁS - outubro 1954 até dezembro 1960. Report. Archive Frederico Waldemar Lange, Box 110

Martins L (1976) Pouvoir et développement économique - formation et evolution des structures politiques au Brésil. Éditions anthropos, Paris

Mello JMC de, Novais FA (1998) Capitalismo tardio e sociabilidade moderna. In: Schwarcz LM (org) História da vida privada no Brasil: contrastes da intimidade contemporânea. Companhia das letras, São Paulo, pp 586–588

Memórias da Bioestratigrafia e da Paleontologia nos 50 anos da Petrobras (2003). Rio de Janeiro: Centro de Pesquisas e Desenvolvimento Leopoldo A. Miguez de Mello

Miranda FCP (1960) Comentários à Constituição de 1946. 3. ed. rev. e ampl. Rio de Janeiro: Borsoi

Moura P, Carneiro FO (1976) Em busca do petróleo brasileiro. Ouro Preto: Fundação Gorceix

Odell PR (1996) Geografia econômica do petróleo. Zahar, Tradução Jairo José Farias Rio de Janeiro

Oliveira Jnior EL de (1959) Ensino Técnico e Desenvolvimento. Ministério da Educação e Cultura, Instituto Superior de Estudos Brasileiros, Rio de Janeiro

Oliveira C (1961) Resumo Histórico do Treinamento na Petrobrás. Boletim Técnico PETROBRÁS, Rio de Janeiro, v 4, n 1–2, pp 71–72, January/June

Organograma básico da Petrobrás (1955) CEnPEs/ Petrobras Library

Passarinho JG [Letter] (1958) Belém [To] LinK, Walter. Relatório da viagem ao rio Moa. 3 f. Archive Frederico Waldemar Lange, Box 108

Pedreira A de B (1927) A pesquiza de petroleo. Typographia do «Annuario do Brasil», Rio de Janeiro

Petrobras vai contratar técnicos francêses para ver se petróleo existe (1961) Jornal Diário de Notícias. 36, n 252, Porto Alegre, 1 Jan

Peyerl D (2010) A trajetória do paleontólogo Frederico Waldemar Lange (1911–1988) e a História das ciências. Thesis. State Ponta Grossa University. Ponta Grossa 116 f

Produção mundial de petróleo bruto – 1955/1960. Revista Petrobras. Archive Frederico Waldemar Lange, Box 32, Rio de Janeiro, ano 7, n 179, 1961, p 1

Seabra O da S (1965) A indústria petroquímica no Brasil. Boletim técnico PETROBRÁS, Archive Frederico Waldemar Lange, Box 32., Rio de Janeiro, v 8, n 1, pp 115–133, January/March

Skidmore TE (2010) Brasil: de Getúlio a Castello (1930–64). Companhia das Letras, São Paulo

Smith PS (1978) Petróleo e política no Brasil Moderno. Artenova, Rio de Janeiro

Távora J (1955) Petróleo para o Brasil. José Olympio: Rio de Janeiro

Um Grande empreendimento econômico lançado no Brasil (1951) Jornal do Brasil, Rio de Janeiro, 5 Dec

Universidade Federal de Ouro Preto - UFOP. Escola de Minas de Ouro Preto. Relação dos Formandos de 1878 a 2007. http://www.semopbh.com.br/arquivos_pdf/livro.pdf. Accessed Aug 2012

Vargas G (1964) A política nacionalista do petróleo no Brasil. Tempo brasileiro, Rio de Janeiro

Vargas M (1994) O início da pesquisa tecnológica no Brasil. in: Vargas, M. (org.). História da técnica e da tecnologia no Brasil. São Paulo: Editora da Universidade Estadual Paulista; Centro Estadual de Educação Tecnológica Paula Souza, pp 211–224

Walter Link deixando o Brasil: 'Cumpri o meu dever' (1961) Revista O Cruzeiro, Rio de Janeiro, ano 21, n 1

4 Improvement, Professionalization, and Geosciences Teaching (1952–1968)

4.1 The Technical Improvement Supervision Sector (SSAT, 1952)

From the end of the nineteenth century, until the thirties, there were initiatives of technical capacitation through Federal Agencies, which were featured by learning through experience (empirical-practical way) and manuals, as previously mentioned. With the discovery of oil in Bahia (1939) and the creation of the National Petroleum Council (1938) and Petrobras (1953), there was a need to improve and to professionalize the workforce. These transformations between technical training, improvement, and professionalization were more evident from the economic and political changes that took place in Brazil at the end of the nineteenth century and throughout the twentieth century.

Science and technology also began to exert more and more influence on the economic growth of the country; according to Manuel Castells (1999), the way of development (organization of technological processes), "technology and technical relations of production are spread throughout the whole set of social relations and structures, penetrating power and experience and modifying them" (Castells 1999: 54), this also happened in the industrial oil sector in Brazil.

The first initiative to improve and professionalize manpower for the oil industry in Brazil began with courses created and carried out both by the CNP and, later, by specialized centers of education at Petrobras, based on existing didactic-pedagogical models with a tradition in the country. As an example, it is worth mentioning that in Brazil, so far as it is known, vocational secondary education combined with entrepreneurship goes back to 1924:

> […] when it was created next to the Viação Férrea Sorocabana, the first railway training center was designed to prepare mechanics, lathe operators, carpenters, boiler workers, blacksmiths, etc. It was an attempt to systematize practical and theoretical teaching by an enterprise, outside the supervision of the official agencies, which already had a network of Industrial, Technical and Higher Schools of Arts and Crafts. The historical merits of Sorocabana are not only the didactic revolution of the use of the methodical series (which allow the rational follow-up of technical-manual operations), but, above all, the establishment of an educational policy in business terms, this way, extending the narrow field of the official initiative, which suffered and suffers from the mitigation of funds and was limited to a routine educational motivation of empirical characteristics (Oliveira 1962: 105–106).

In the thirties, the Institute of Technological Research (IPT), an institution that offered important contributions to the country industrialization process (Salles-Filho 2000), building sections for soil and geology research, and contributed to the subsequent structuring of activities related to the courses offered by CNP and Petrobras. The IPT was also responsible for setting up and conducting the first courses in Soil Mechanics in the forties (Futai, s.d.).

Other examples of technical training centers that contributed to both the CNP and Petrobras were: the National Industrial Learning Service

D. Peyerl, *The Oil of Brazil*, Historical Geography and Geosciences,
https://doi.org/10.1007/978-3-030-13884-4_4

(SENAI) in 1942 designed to train able and qualified professionals for the nascent base industry; the Intensive Industrial Labor Preparation Program (PIPMOI),[1] created by the Industrial Education Board of the Ministry of Education and Culture in 1963; the Technological Institute of Aeronautics (ITA), which participated mainly in the maintenance of refinery equipment in 1950.

It is also noteworthy, in another instance, the wit of Robert Cochrane Simonsen (1889–1948)[2] who gathered "around the idea of preparing the ranks of qualified workers, a powerful group of industrialists, awakening them to the meaning that labor has as a factor of productivity" (Oliveira 1962: 106). In the forties, Simonsen proposed a planning of the Brazilian economy that focused on the structuring of education, based on a system of technological research and on the training/intensification of professional education (Federation of … 1941), as well as the creation of new engineering schools and the dissemination of institutes of technological, industrial, and agricultural research. In the period, his ideas, for political reasons, were not put into practice, but they represented incentives and investments in the improvement and professionalization of labor in Brazil. Only in 1968, through the Strategic Development Program (PED 1967–1970),[3] the idea of Simonsen to incorporate a scientific-technological policy into global economic planning was eventually adopted (Ferreira 1983).

Thus, the great step toward the improvement and professionalization of the Brazilian manpower necessary for the exploration, exploitation, and production of oil occurred in 1952. This initiative was taken by the National Petroleum Council when structuring the Technical Improvement Supervision Sector (SSAT) with the purpose of generating specialized and technical manpower. The first measure coordinated by SSAT occurred in 1952, with the creation of the first Oil Refining Course. In the same period, two specialists in the area of Petroleum Geology, Leverson,[4] and Ducan Macnaughton[5] were invited by the CNP to try to solve the manpower problem, first suggesting the installation of a Course in Petroleum Geology (Entrevista … 1982). However, the idea did not go ahead because of the concentration, in the period, in the construction and the operation of refineries.

For the creation and validation of the Petroleum Refining Course, the SSAT resorted to an agreement with the University of Brazil, through the National School of Chemistry. The course proved in practice the fulfillment of the main objective of SSAT: the improvement of manpower in training professionals to work in the refineries.

> The requirements of the first two state refineries – in Mataripe and Cubatao – imposed absolute priority on the formation of teams of technicians in industrial processes that would allow the country to be exempt from heavy burden with the hiring of foreign technicians, resulting in the inauguration, in 1952, of the Oil Refining Course, officially recognized as a university extension program […] (Moggi 1968: 1).

The initiative considered promising good results obtained by the first Oil Refining Course,[6]

[1]"The program had as its initial plan to operate for 20 months but was maintained for 19 years. Its qualification activities began in 1964, executing governmental projects until the year of 1982. It had as a motto to respond to the growth of the industrial park in Brazil with training of labor" (Machado and Garcia 2013: 48).

[2]Graduated in Engineering by the Polytechnic School of São Paulo. He acted as a politician, sociologist, and professor (Economic History of Brazil). He was the author of several books, focusing mainly on economic aspects.

[3]"[…] the PED proposed to act directly and indirectly to increase liquidity, reduce cost pressure and increase demand, by reducing the pressure that the public sector exerted on productive activity. With this, the government intended to achieve the two basic objectives of this plan: accelerating growth and containing inflation" (Rezende 2010: 56).

[4]No bibliographic references were found.

[5]He studied geology at the University of California, where he became a Geology professor.

[6]"The first ceremony for the end of classes and certificates of specialization in petroleum refining was held on 30 June 1953, and only nine technicians among industrial chemists and engineers obtained the diploma: Tarcisio Barroso, Alfredo Ferraz, Ivan Sá Motta, Ivo de Souza Ribeiro, Ilena Horta Zander, Alberto Boyadijan, Gloria Conceição Klein, Haylson Oddone and José Angrisani" (Caldas 2005: 31).

ended up having as an ally the rapid development of the oil and refining industry in Brazil, which was increasingly focused on energy issues of raw materials. The Oil Refining Course taught:

> [...] together with equipment processing and maintenance techniques, the first graduated teams competed to even raise the load of the Landulpho Alves Refinery to 10,000 barrels per day. It is easy to explain why refining deserved priority over other sectors. It constitutes an industrial operation of certain profit and reduced risks, not to say null (Oliveira 1961: 141).

The objective is to train people "with sufficient technical background so that they could not only cook with the recipe of cake given to them" (Moggi 1988: 128). It was, therefore, necessary to learn, to incorporate, to create, and to produce the technology itself, and not merely to reproduce it through manuals.

The main difficulty for the creation of the courses and even for their continuity—was the scarce resources destined for the CNP, more so for a newly created sector such as SSAT. With great effort, according to the reports of the protagonists, the first course of improvement was carried out with an average duration of one year.

The SSAT was headed by the engineer Antonio Seabra Moggi,[7] who held the position until the absorption of this sector by Petrobras in August 1955 (Caldas 2005). Moggi is considered, along with other names, as Leopoldo Américo Miguez de Mello (Moggi 1988), one of the great idealizers of the creation of the courses of improvement and professionalization and the construction of centers geared to the teaching and research of geosciences in the country. The very words of Antonio Moggi express this well:

> [..] we thought we would take a big step if we could train high-level technicians, not only that they could operate refineries - just cooking third-party recipes - but even make their own recipe, that is, design a unit, understand how it can be designed and know what we call the "black cover book", that is, entering the closed box and unrolling it, being able to understand how it is manufactured (Moggi 1988: 101).

By 1955, "when there were already four classes formed, the technical improvement supervisory sector was extinguished, and all its assets transferred to Petrobrás" (Moggi 1988: 130). This absorption was favorable to the continuation of the work started by SSA, since the company resources for further training and professional training increased considerably, especially when it was convinced and prioritized the problem of staff training.

In this period, the technological development of the oil industry in Brazil began, which changed its professional and technical framework through government support. Although the country had been conducting oil research since the beginning of the century through government agencies, knowledge of the techniques used for oil prospecting and/or refining was dependent on foreign technology and technique: "there was no deep know-how" (Moggi 1988: 67).

It should be noted that the manuals used in the application of the technique were in English or French and that many terms were interpreted in a random way, therefore, another didactic-pedagogical policy of the courses, both the CNP and, later, of Petrobras, was to have English as a basic subject. This was also important because most of the foreign teachers had their classes in English. From this point on, the need to develop their own know-how was created through the creation of sectors and centers that could develop techniques, equipment and, above all, to train Brazilian professionals who could learn and deepen their knowledge related to oil.

[7]Born in Italy, on 20 December 1920. He entered the National School of Chemistry in 1941, and when he finished his studies in 1944, he went to the United States, where he completed his degree in Chemical Engineering at Vanderbilt University. He began his career at the National Petroleum Council (CNP) in 1947 as a cabinet officer for the president of the CNP, General João Carlos Barreto, and participated in the Commission for the Constitution of the *Refinaria Nacional de Petróleo S.A.* He was fully involved in the creation of specialization courses in Geology by the Center for Improvement and Petroleum Research (CENAP) and courses at the universities. He was superintendent of CENAP from 1955 to 1962 and from 1964 to 1965, as well as superintendent of the R&D Center Leopoldo A. Miguez de Mello from 1966 to 1980.

4.2 The Oil Improvement and Research Center (CENAP/Petrobras, 1955–1966)

In 1955, the Oil Improvement and Research Center (CENAP) was created by Petrobras, with a comprehensive program that aimed at the preparation of specialized labor, necessary for the expansion of activities related to the exploration and industrialization of oil reserves (Fortes 2003). The need for specialized labor became indispensable in this period because of the "transition from an essentially agricultural Brazil to an urban, industrial, and service Brazil between 1950 and 1980" (Fausto 1995: 539), as well as the creation of an industrial complex such as Petrobras.

The oil industry needed short-term manpower. Since the beginning of the twentieth century, it was already possible to observe this need, but it always appeared in a second plan due to the agro-exporting economy that marked Brazil and the fact that oil was found at a commercial level only in 1941, in Candeias (Bahia State), as already mentioned.

According to the initial survey conducted by Petrobras itself in the mid-fifties, it was estimated that the greatest need for professionals was related to geology, production engineering and drilling and refining engineering. One of CENAP main objectives was then focused on training teams of Brazilian professionals, who should gradually replace foreign professionals. This did not happen immediately, as it was demonstrated in the second part of this book, but the idea of replacement remained and somehow strengthened the advances in teaching and in-country improvement/training.

These advances were consolidated on 22 April 1957, when the company Executive Board, formed by Janary Gentil Nunes (President of Petrobras),[8] Irnack Carvalho do Amaral (Director of Operations),[9] João Tavares Neiva de Figueiredo (Economic-Financial Director)[10] and José de Nazaré Teixeira Dias (Director of Administration),[11] approved and transmitted, through Resolution No. 7/57, the guidelines for the Training and Improvement of Personnel Plan, approved by the Petrobras National Board of Directors, for the knowledge of the Petrobras units. The purpose was to provide Petrobras with the specialized personnel that the company needed to fulfill the work programs (Petrobras 1957d).

The resolution emphasized the need to "overcome, by the most appropriate means and strictly necessary deadlines, the shortcomings of the education system of the country in terms of personnel training for the oil industry" (Petrobras 1957d). Thus, the priority was to hold "special courses, internships, roundtables at work, round tables, scholarship grants for study at home and abroad, and publications" (Petrobras 1957d). The resolution still had some requirements, such as in the case of an internship abroad, where all applicants for further training abroad should have prior experience in the company services so that the technique used abroad would be properly and better applied here.

Other points, resulting from the guidelines proposed in 1957, resulted in the subdivision of CENAP in the following groups: Senior Technical Personnel, Higher Level Personnel, Administrative Personnel, Maritime Personnel,[12] Brazilian Institute of Petroleum and Complementary Activities (namely, Documentation,

[8]Born in Alenquer, Pará State, on 01 June 1912. In 1956, he assumed the presidency of Petrobras, where he remained until 9 December 1958. In 1962, he became a federal deputy for the Amapá State, and later he started to dedicate to the private initiative. He passed away in Rio de Janeiro, on 15 October 1984.

[9]Born in Rio de Janeiro City on 6 October 1905. Graduated in Engineering in 1931 by the EMOP. He worked as a SGMB engineer, was a special assistant to the president of CNP, headed the Geophysics section of the Division of Mineral Production and was a director of Petrobras twice, from 1954 to 1957 and from 1961 to 1963. He passed away in Rio de Janeiro City, on 8 January 1983.

[10]Graduated in Mining and Civil Engineering by the EMOP in 1936. From 1942 he worked as a geologist in the Division of Mineral Production and Petrobras.

[11]For more information, see Dias (1991).

[12]"In the preparation and improvement of the maritime personnel, cooperation will be sought with the Merchant Marine School of Rio de Janeiro and with the other specialized courses of the Ministry of the Navy and private entities" (Petrobras 1957c: 1).

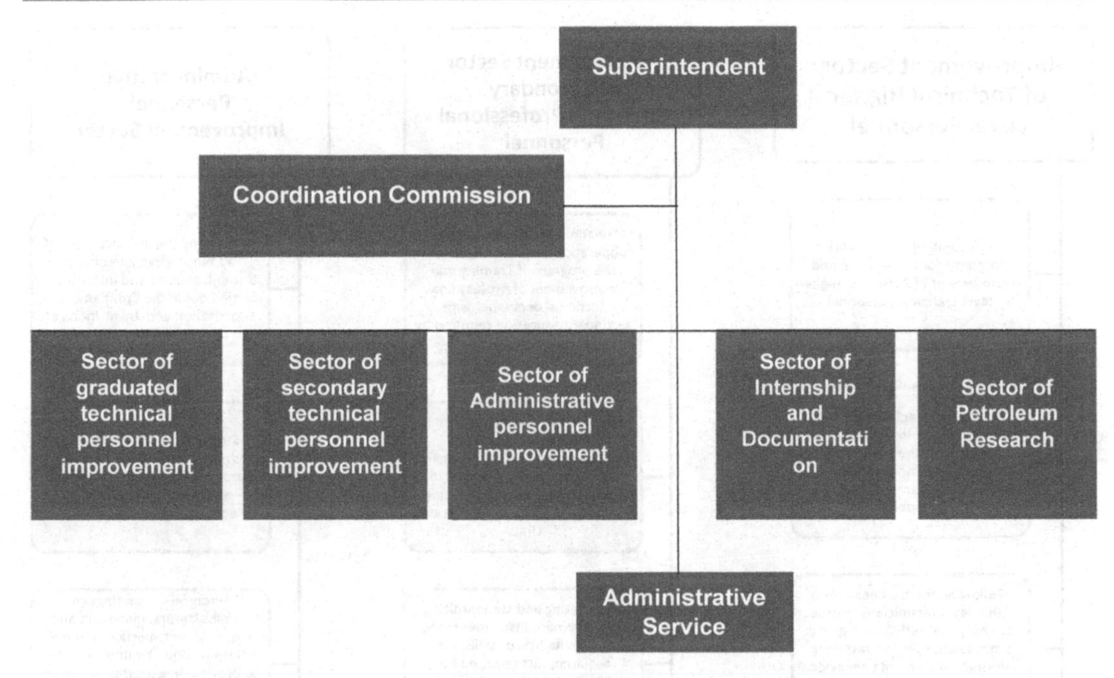

Organizational Chart 4.1 CENAP organization (1957) (Information from Petrobras (1957c))

Studies and Exchange, Work Safety, Knowledge of Foreign Languages). The CENAP structure was organized as it follows: of the CENAP organization (Organizational Chart 4.1), three sectors deserve our attention: (1) Higher Level Technical Personnel improvement sector; (2) Secondary Education Professional improvement[13] sector; and (3) Administrative staff sector (see Organizational Chart 4.2, p. 161).

In addition to the aforementioned points, we opened a parenthesis for the Petroleum Research sector, which aimed to coordinate and supervise the studies and research conducted by the working groups, which were set up with the objective of promoting the development of scientific knowledge and Petroleum technology, and directly execute the laboratory research and studies attributed to CENAP (Petrobras 1957c).

The CENAP also opted to invest in the expansion and training of professionals, bringing high-level teachers from abroad (Caldas 2005: 11). CENAP had as one of the scopes to be fulfilled, to provide, by the most appropriate means and strictly necessary deadlines, the shortcomings of the country educational system with respect to the training of personnel for the oil industry. In addition, it was also up to the Brazilian technicians to have the opportunity to familiarize themselves with the techniques and know-how of foreign specialists at the service of Petrobras (Petrobras 1957c).

Through the CENAP, several courses were offered, among them: Refining Course; Course of Petroleum Geology; Drilling and Production Course; Special Petroleum Course; Course of Maintenance of Petroleum Equipment; Special Maintenance Program; and Petroleum Engineering Course, among others. The selections of students covered a number considered satisfactory by the company staff so that the first courses had a total of selected candidates, of whom a few hundred were recruited (Moggi 1988: 131). Most

[13]The use of the word "improvement" at this time differs from the initial definition pointed out in this part, generalizing it to the Higher and Secondary Degree. We added that at the moment, and even in the use of some authors, there was no such differentiation, so it will be possible to find, during the text, placements as pointed out now.

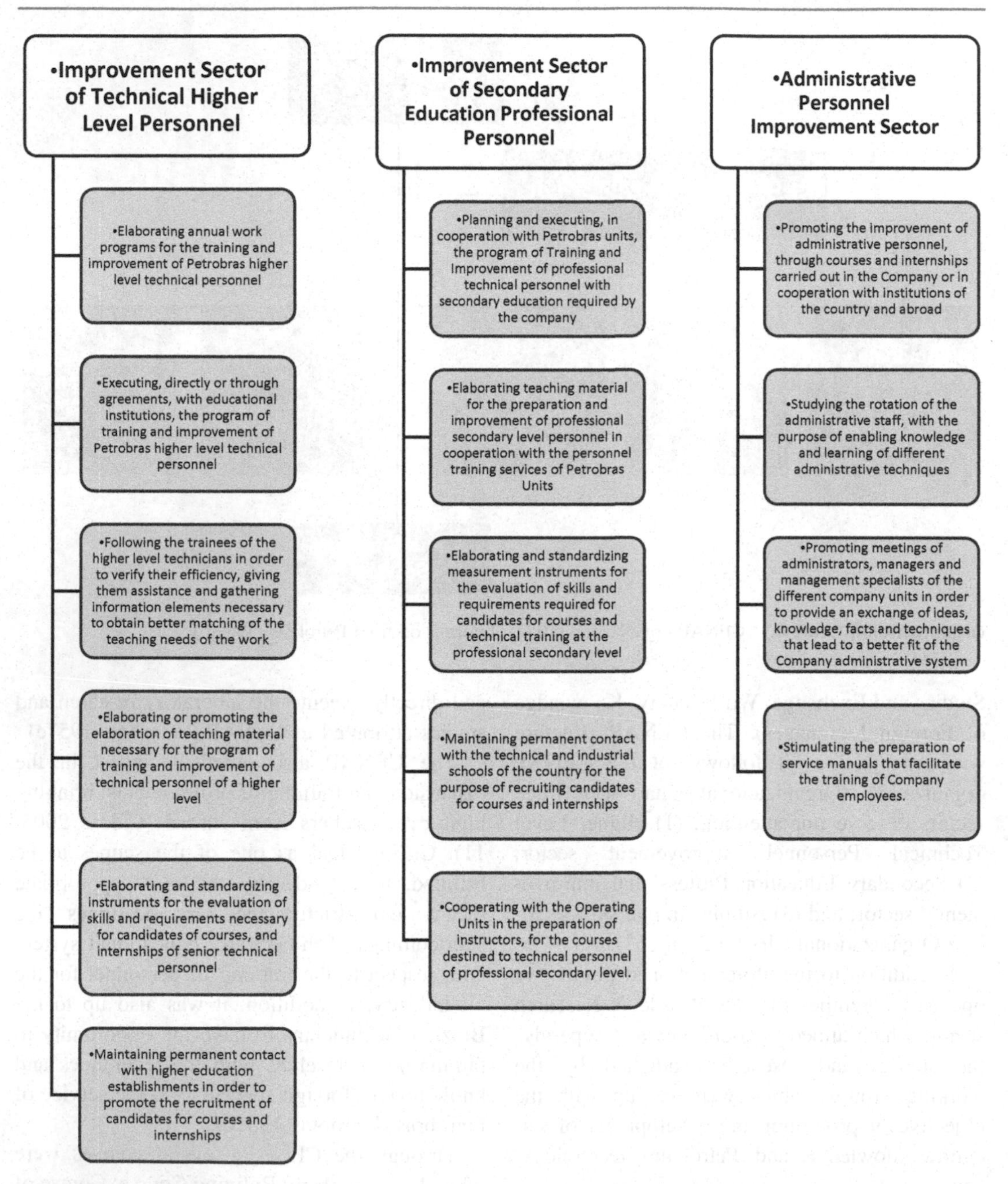

Organizational Chart 4.2 Description of three sectors of CENAP (Information from Petrobras (1957c)). *Source* Elaborated by the author

courses were held at Petrobras dependencies in states where there were regional centers or departments such as Bahia and Rio de Janeiro.

The selection and recruitment by the CENAP for higher level courses took place at the national level, being applied in those states that had universities and that had suitable candidates for the vacancies. This process was disclosed as follows: (a) through manuals prepared and published by CENAP/Petrobras; (b) through the press (mainly by the newspapers); (c) through correspondence to the universities; and (d) by sending of groups of technicians and professors who went from university to university, "exposing what CENAP courses were, attracting the candidates" (Moggi 1988: 131).

In August 1956, for the first selection, CENAP:

> [...] contacted 16 US companies related to the oil industry, asking for information about the measures used in the selection of technical personnel to train and work in these companies. CENAP received, from most of them, test models and interview records used for selection purposes. It has been checked after studying this material that it would be necessary to organize a battery of tests from Petrobrás selection, which could select engineers capable of becoming good technicians (Recrutamento e ... 1957: 59).

After recruitment, students were subjected to selection tests composed of tests of scientific knowledge (weight 3: 60 questions of physics, chemistry, natural history, and mathematics), general aptitudes (weight 2: vocabulary test, numerical test and shape/graphs tests), and English proficiency (weight 1: English comprehension) (Recrutamento e ... 1957: 59).

Between 1956 and early 1957, the selection was made, applied in a total of 30 universities/schools/colleges all over Brazil (Table 4.1). It should be noted that, in table below, we did not find the name of the School of Mines of Ouro Preto, suggesting that there were no subscribers in this period. However, we recall that many mining and civil engineers named in this book who worked at Petrobras belonged to EMOP.

In this period, Brazilian universities did not yet have courses geared to the area of petroleum or the even a degree in geology—which was officially created in 1957. Most of the candidates were then civil engineers and mines or even the

Table 4.1 Selection of candidates to CENAP courses—1957 (Data were partially taken from *Recrutamento e ...* (1957: 60))

Schools	Subscribed	Present	Completed the tests	Approved	Not approved
School of Engineering of the Mackenzie University—São Paulo State	89	51	43	23	20
Polytechnic School of the University of São Paulo—São Paulo State	62	39	39	30	9
National School of Engineering of the University of Brazil—Distrito Federal	59	52	52	38	14
Engineering School of the University of Minas Gerais General—Minas Gerais State	45	7[a]	35	24	11
School of Engineering of the University of Paraná—Paraná State	40	20	19	12	7
Institute of Aeronautical Technology—São Paulo	24	19	16	15	1
School of Engineering of the University of Rio Grande of South—Rio Grande do Sul State	20	7	7	6	1
School of Engineering of Juiz de Fora—Minas Gerais State	20	18	17	13	4
Institute of Electrotechnical Itajubá—Minas Gerais State	20	18	18	12	6
Polytechnic School of the Catholic University of Rio of January—Rio de Janeiro State	19	12	12	10	2
Faculty of Industrial Engineering—São Paulo State	16	11	7	6	1
School of Engineering of the University of Recife—Pernambuco State	14	13	12	8	4
School of Engineering of the University of Pará—Pará State	13	8	8	1	7
School of Chemistry, University of Paraná—Paraná State	12	6	6	3	3
Polytechnic School of the Catholic University of Pernambuco—Pernambuco State	9	9	9	4	5

(continued)

Table 4.1 (continued)

Schools	Subscribed	Present	Completed the tests	Approved	Not approved
Fluminense Engineering School—Rio de Janeiro	6	6	6	5	1
Faculty of Philosophy, Sciences and Arts of University São Paulo—São Paulo State	6	3	3	1	2
Faculdade de Filosofia, Ciências e Letras da University de Curitiba—Paraná	5	4	4	–	4
Faculty of Philosophy, University of Ceará—Ceará State	4	4	4	–	4
Faculty of Philosophy, Sciences and Arts of Paraná—Paraná State	4	4	3	2	2
National School of Chemistry of the University of Brazil—Distrito Federal	3	3	1	2	1
School of Chemistry of Sergipe—Sergipe State	3	1	3	–	1
Faculty of Philosophy of Pernambuco, University of Recife—Pernambuco State	3	3	2	1	2
Faculty of Philosophy, University Catholic of Pernambuco—Pernambuco State	3	2	–	2	–
School of Chemistry, University of Recife—Pernambuco State	2	–	–	–	–
Faculty of Philosophy, Sciences and Letters of University in Campinas—São Paulo State	2	1	1	1	–
Faculty of Philosophy, Sciences and Letters of Univ. Mackenzie—São Paulo State	2	2	2	1	1
Faculty of Philosophy, Sciences and Arts of University of Distrito Federal	1	–	–	–	–
Faculty of Philosophy of the Pontifical University Catholic of Rio Grande do Sul—Rio Grande do Sul State	1	–	–	–	–
Faculty of Philosophy, University of Brazil—Distrito Federal	1	1	1	1	–
Candidates who did not report the school they attended	9	5	4	1	3
Candidates who did not fill the register	5	1	1	–	1
30 Schools/Universities—TOTAL	522	360	339	222	117

Source Elaborated by the author
[a]It is believed that the corresponding number would be 37

area of natural history, while the need, in 1957, was 130 professionals for the refining, geology and maintenance areas, mainly.

For a comparison purpose, it is noteworthy that after two years, in December 1959, in another selection, a total of 532 candidates from different Brazilian states subscribed to the tests to enter the following specialization courses kept by Petrobras. Petroleum Geology, Maintenance of Petroleum Equipment, Drilling and Oil Production (Interesse dos … 1959) (Fig. 4.1).

The tests made by the same visited cities, in October 1959, by the technicians' teams from Petrobras.

They were specially prepared for the recruitment of skilled manpower for the oil industry (Interesse dos … 1959) (Table 4.2).

The teaching offered by CENAP was based on a didactic-pedagogical system based on the university system, with characteristics such as attendance in practical and theoretical classes; stages; faculty formed by renowned professionals; curricular matrices based on other universities (national and foreign, considering, above all, engineering); and courses taught by professionals who know the area. The reprobations were not tolerable (Caldas 2005). The aim of the CENAP was also to "stimulate the improvement of the teaching of the basic

Fig. 4.1 Test in the Engineering School from the University of Recife—December 1959. *Source* Interesse dos … (1959: 7)

Table 4.2 Number of candidates subscribed for the selection of 1959

City	Number of candidates per city
Recife	84
Distrito Federal and Niterói	75
Belo Horizonte	66
Curitiba	54
São Paulo (capital)	36
Porto Alegre	34
Belém	28
Salvador	22
Itajuba	20
Itaguaí	20
Fortaleza	17
Juíz de Fora	11
São Jose dos Campos	11
Piracicaba	9
Vitória	8
Pelotas	8
Aracaju	7
Maceió	6
Sao Carlos	6
Cruz das Almas	6
Campina Grande	4

Source Interesse dos … 1959

sciences and the sciences of the profession in technical–scientific courses, including through agreements with universities and institutes of higher education" (Moggi 1961: 1). Thus, Petrobras gradually transferred training and staff development programs to universities and secondary schools (Moggi 1961), seeking to maintain articulation with the country educational system, according to its stated intention:

> In compliance with the terms of an agreement negotiated by CENAP – in 1959 between Petrobras and the national industrial learning service – SENAI, two Vocational Training Centers (CEFAT) were built, one in Candeias City, Bahia State (fifties), and another in Cubatão City, São Paulo State (sixties). Since its inauguration, they have basically functioned as Vocational Technical Schools and provided industrial learning for minors, children of

> employees and the community; professional improvement for employees and, in addition, basic schooling. SENAI and Petrobras agreed that the Centers could be used, as supplementary, as training agencies for local unit employees (Caldas 2005: 33–34).

We have made a parenthesis here to mention that, in fact, Petrobras contributed to the opening, in Brazil, of several courses—mainly Geology. This process has continued with the creation of new courses according to the demands and studies directed to the research of oil in the continental shelf (1968). The case that we can more specifically point to is the creation of the Petrobras University in 2004, which has since continued to train the professionals of this company.

In September 1959, the CENAP was further differentiated by organizing a Research Seminar, which included "representatives of the Presidente Bernardes Refinery, Fertilizer Factory, Bahia Production Refinery, Shale Industrialization Superintendence, assistance of Refining" (Williams 1960: 161), as well as CENAP own technicians. The consolidation of the seminar aimed in establishing and defining internal priorities for CENAP for 1960, namely:

1. **Research Applied in Refining and Petrochemistry**: carried in four pilot-units: Catalytic Cracking Fluid Midget FCC; Catalytic reforming; distillation—capable of distilling batches of 318 L of oil or other material; and Continuous Extraction, type Rotating Disc Contactor, being able to operate with a very wide range of solvents, such as propane, phenol and furfural, of great interest for the study of lubricating oils.
2. **Research in Exploration and Production**: Collaboration with the Geology course, assisting in the correlation of geological data on the different sedimentary basins of Brazil.
3. **General Interest Research**: Oil analysis and evaluation; studies and analyses of gases; studies and analyses of catalysts; and general studies.
4. **Organization of Technical Manuals**: Elaboration of the Manual of Technical Data, of the Manual of Projects of Processing, in addition to the organization of the research seminars (Williams 1960).

In addition to the organization of seminars, CENAP also invested and extended its improvement potential to the research and development phase in postgraduate courses (Williams 1960). It should be noted that CENAP continued to conduct the courses in partnership with other universities.

In 1961, Antonio Seabra Moggi elaborated the guidelines for the Perfection and Professionalization of Petrobras Personnel. From these guidelines, it was up to Petrobras to provide, by the most proper means and by the strictly necessary deadlines, the shortcomings of the country educational system, with respect to the improvement and professionalization of personnel for the petroleum industry (Moggi 1961). Somehow, Petrobras tried not to restrict the investment of education initiated within the company but encouraged it to expand through universities and secondary and technical education.

Although CENAP has not been able to reach its degree of self-sufficiency in terms of trained personnel, due to the increased demand of professionals in the area of geosciences, its initiatives and those of SSAT have contributed and modified the direction of improvement and professionalism, demonstrating that it was possible for Brazil to invest and build part of its know-how, even though it would have been easier to import technology and labor from abroad. The nationalism of this period was essential to invest in the formation of Brazilian labor:

> In other words, the time spent in the assembly of an industrial plant (refinery, asphalt plant, synthetic rubber, fertilizers, etc.) or in the construction of an element of transportation of crude and derivatives (oil tanker, maritime terminal, petroleum, etc.) is shorter, using foreign engineering rather than preparing the national workforce capable of operating them without dependence on any kind of their original planners and designers (Oliveira 1961: 71–72).

The country, fortunately (looking at the results afterwards), managed to build its own know-how, and these changes were perceptible, especially since the revolution of 1930, the establishment of the New State (which guided pedagogical and didactic presuppositions based on nationalistic, disciplinary, moral and professional values) (Oliveira et al. 1982: 154) and the Juscelino Kubitschek Government (1956–1961), which contributed to the economic and political development based on construction and technological development and the preparation of the national workforce. It was not only higher education professionals that CNP and Petrobras needed, but a professional staff focused on consolidating the oil industry.

At first, the concern was to specialize in higher education professionals, then Petrobras expanded its plans for technical and secondary-level positions (electronics technicians, seismic teams, drilling supervisors, etc.). For reasons already mentioned, the influence that SENAI exercised caused a backlog of know-how, giving CENAP the use of its methodological structure, which was based on it to outline the planning of Petrobras mid-level scenario (Oliveira 1961: 106).

Thus, the work carried out by CENAP, industrial mobilization, technological development and the creation of new university courses led to the expansion of the petroleum industry to research-centered development. Petrobras invested in the creation of a Research Center, seeking to solve technical and, in parallel, labor problems through scientific, technological and practical knowledge.

4.3 Research and Development Center Leopoldo Américo Miguez de Mello (CENPES, 1963)[14]

At the end of 1960, a working group was created[15] by Petrobras itself, designed to study the creation of an agency that could conduct technological research activities for the petroleum industry (Petrobras 1960). This group was responsible for the following:

(a) the setting of the objectives of the agency, with the clear delimitation of the scope of its activity regarding the various levels of research and technological development;
(b) the recommendation of the internal structure of this organization, its subordination to the direction of the company and its relations with the other sectors and units of the organization;
(c) the recommendation of the preferential location, the indispensable area, the initial staff and the facilities required for the first years of its operation;
(d) the presentation of an estimate of annual expenditure, considering the works that should be executed during the next 5 years of the creation of the agency. This estimate should also include expenditure on the functioning of the agency, if possible; and
(e) the means of action and articulation with the other units of the company and with the technological and scientific research and educational institutions of the country and from abroad (Petrobras 1960).

In this way, the purpose of the Research Center was to carry out research of scientific or technological interest for the oil industry, and it was also up to it to collect, systematize and disseminate documentation of scientific or technological interest to the oil industry (Organizational Chart 4.3).

For this organization, which was to be set up in 1966, a number of foreign companies (Organizational Chart 4.4) were contacted and visited that could, in some way, contribute to the constitution of this agency. According to the companies responses, information was gathered, and a proposal was made for the creation and organization of the Research Center.

One of the main issues that led to the creation of the Research Center was the need to have a single site for research, covering the entire production chain, i.e., exploration, production, refining, and petrochemicals. The site for the installation of the Research Center should be close to a large university—initially suggested in São Paulo City or Rio de Janeiro City—to facilitate technical recruitment, scientific exchange, and research cooperation. In order to achieve this

[14]CENPES was created in 1963, but its activities started only in 1966. In some primary sources, information that establishes the CENPES creation in 1966 can be found, due to the beginning of its activities.

[15]There is no direct information of whom were part of this group.

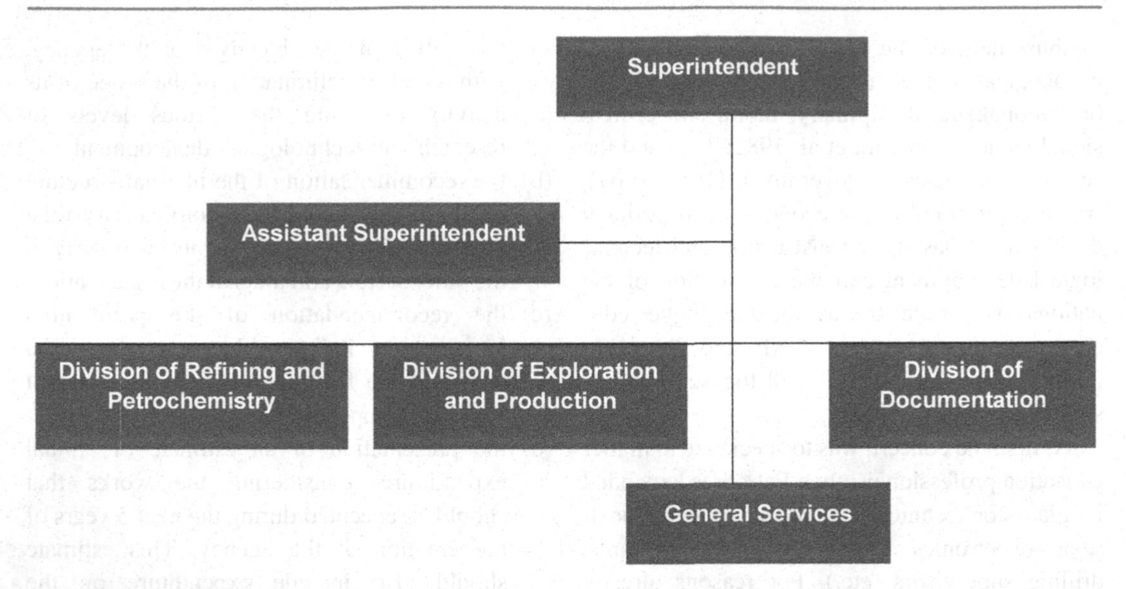

Organizational Chart 4.3 Organization of the Research Center. *Source* Petrobras (1960)

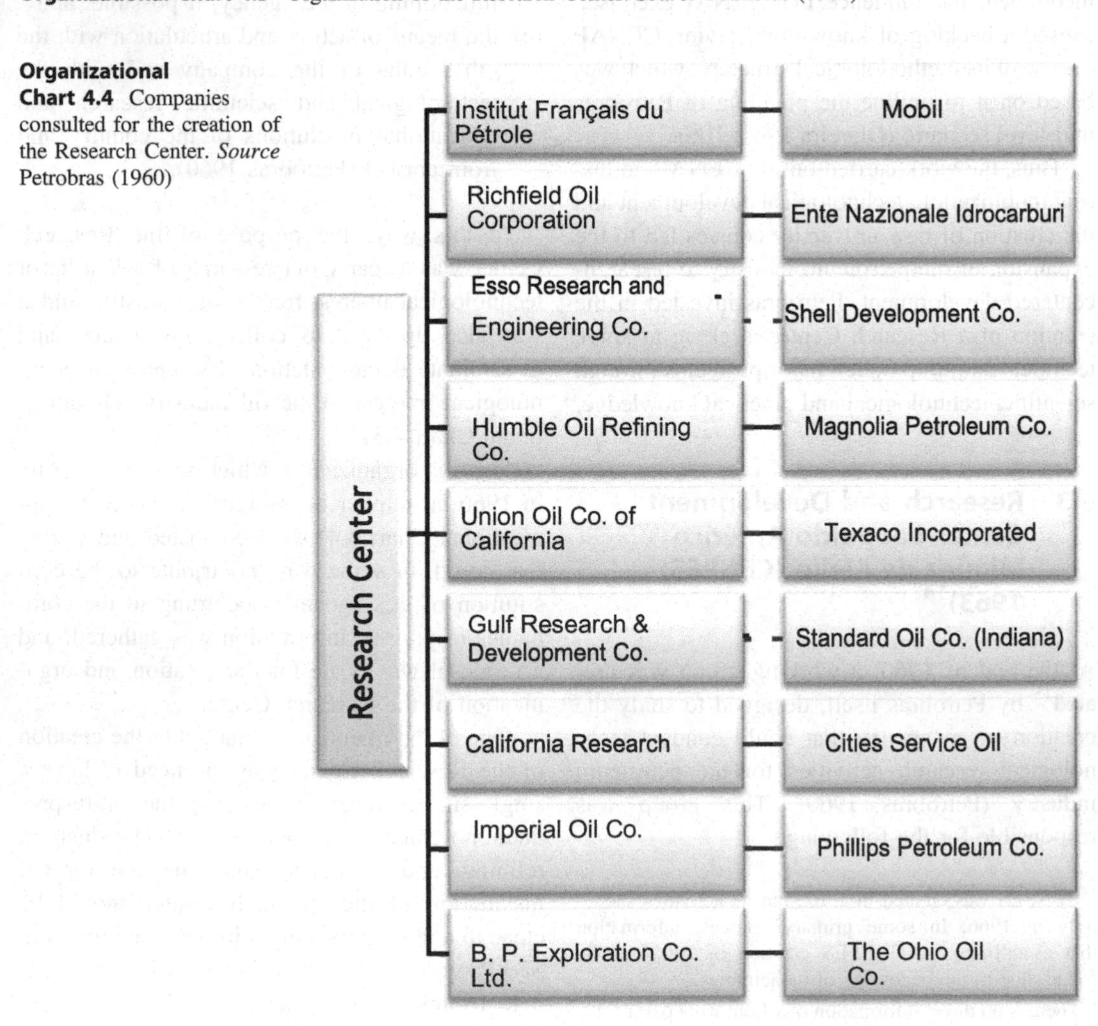

Organizational Chart 4.4 Companies consulted for the creation of the Research Center. *Source* Petrobras (1960)

decision, the problem of the proximity of the Research Center to the operating units was put into question. Based on the responses of the foreign companies consulted, it was concluded that it would be better to move farther from the operating units and, on the other hand, to the proximity of a more university environment.

The main justifications for the construction of the Development Center were centered on the reduction of operating costs, the reduction of the exchange rate, the better performance of equipment or processes and the expansion of the market due to the discovery of new products or new applications of products (Petrobras 1960).

In relation to the improvement and professionalization activity, the Research Center would cooperate in the execution of a postgraduate program, being limited to the administration of teaching or to the follow-up of internships in its laboratories. Initially, it would contribute to the company's technical services seeking ways to improve the techniques used and developing methodologies for them. It would also interfere with the company's operational services, under different angles and purpose of the CENAP, starting to undertake independent studies, mainly in the field of exploration/production.

From the points raised by this group and its improvement, over the years, on 1 January 1966, CENAP is extinguished and the activities of the Research and Development Center Leopoldo Américo Miguez de Mello (CENPES) are initiated, with different perspectives from CENAP and according to new demands that arose in the period but based on the guidelines of the group formed in 1960 for this purpose.

In 1967,[16] the definition of CENPES was announced and the agreement with the Federal University of Rio de Janeiro (UFRJ), signed on 14 March 1968. CENPES absorbed CENAP in the areas of Refining and Petrochemical Research and exchange and documentation.

In the primary sources collected and articles that provide information about CENPES, it is clear the use of applied research as a methodology for forming staff. This is mainly because the division between basic science and applied research has already become institutionalized at a global level (Beer and Lewis 1963), and rapid and cost-effective results have been achieved through scientific and technical processes. In the view of the company, the choice of this option was due to the need for a practical nature of the work, while at the same time proposing to improve and to adapt the specific conditions and procedures already in place (Gubler 1967):

> Applied research can also be defined as the application of all existing knowledge in the practical solution of a specific problem. This conception results in the need for proper team capable of applying ≪all this existing knowledge≫. For ≪proper personnel≫ it is understood employees in sufficient numbers, having diversified training at high competence. For research, it is necessary to look for technically creative personnel. Despite the variety and complexity of the equipment, a department will cease to be a research unit if it does not rely on scientifically creative people (Williams 1960: 161–162).

One of the difficulties encountered, at the beginning of the CENPES activities, according to the reports, was the "lack of experience of the young people and the insufficiency of the secondary-level professional workforce to perform certain routine work entrusted to technicians of university level, resulting in a sub-employment of elements of value" (Gubler 1967: 8). In terms of research, CENPES focused their activities in some specific studies, such as stratigraphic and basin studies (dating, rock texture, mineralogical composition, and organic content); paleontological studies; geochemical studies; particular surface geological studies; geophysical, technological, and mathematical studies (seismic, mathematical, gravimetry, electro-resistivity); production and drilling studies (reservoir geological studies), technical and scientific documentation studies; and particular economic studies (Gubler 1967). Some of the recommendations focused on immediate needs, such as design of general stratigraphic studies of the basins; geophysical studies projects; seminar on seismic; information on gravimetry; information journeys on terrestrial magnetism;

[16] In this period, Petrobras already relied on 36,048 employees.

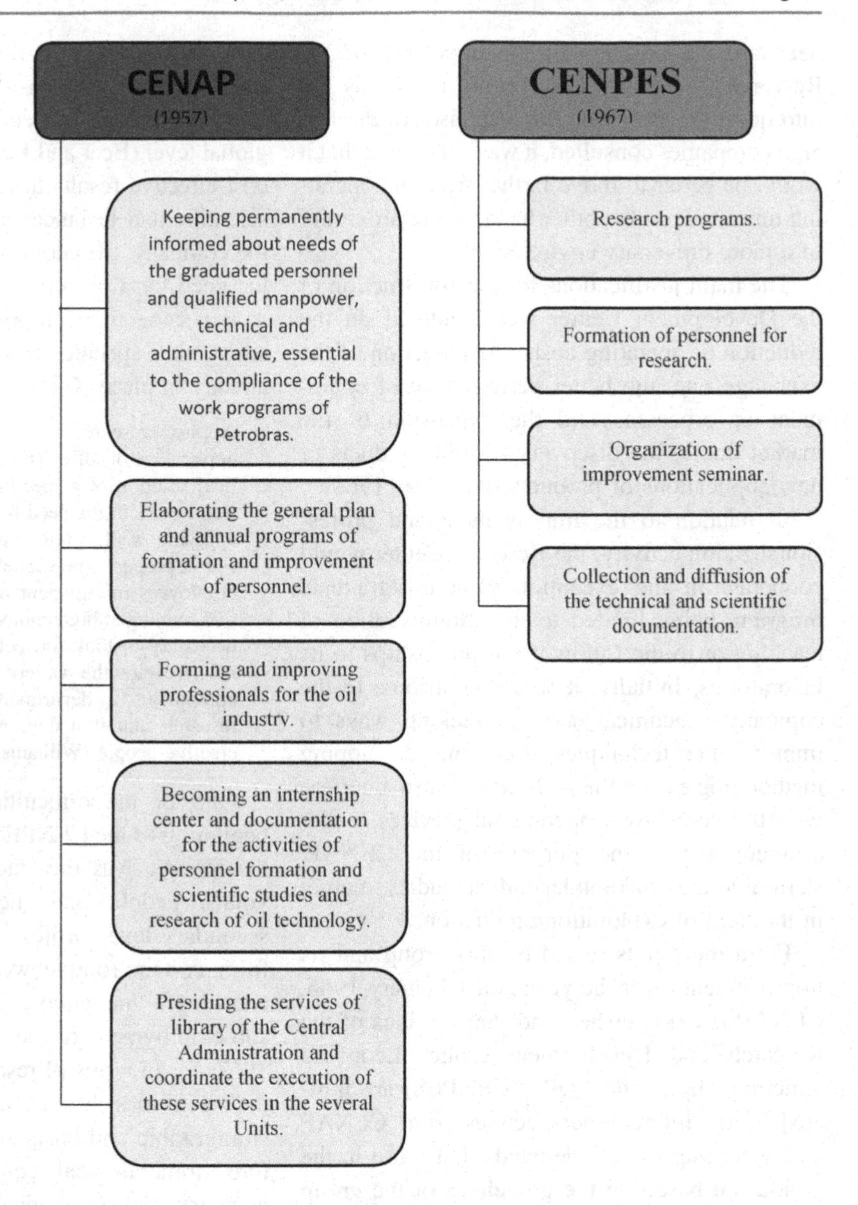

Organizational Chart 4.5 Main objectives of CENAP and CENPES. *Source* Petrobras (1957c), Gubler (1967)

information journeys on basalts and dolerites; projects of production and drilling studies, seminars on rock fracturing, seminars on secondary recovery, and information journeys on specific subjects (Gubler 1967).

As pointed out by Marc Albo,[17] CENPES needs and technological development also pointed to urgent demands, already pointed out during CENAP existence, such as in the field of geophysics, the need for technicians not totally laymen in the area and certainly highly specialized in mathematics, physics and advanced electronics—for this reason, usable in the various branches of geophysics; the need for consultations and agreements with foreign organisms and a certain correlation with consultations with other techniques; and the need for national technicians.

Despite the clear differences (see Organizational Chart 4.5), CENPES continued with some activities developed and worked by CENAP,

[17]No biographical information was found about Marc Albo. He is the author of records from Petrobras related to CENPES.

providing the training of professionals for research through the proposal of an annual program of improvement and professionalization in the country and abroad, and giving continuity of the organization of seminars, such as those carried out by CENAP and already mentioned. That way, in addition to contributing to the improvement and professionalization, the initiatives of SSAT, CENAP, and later the CENPES eventually formed and brought together a technical–scientific network composed by, as Luís Medina Peña defines, political actors (technicians, petroleum engineers, geologists, etc.), that is, a group of professionals with scientific training, who, through their actions, obtain effects on policies and influence the organization of the modern state (Medina Peña 2004). This definition, in our opinion, can summarize the role that CNP and Petrobras played in this process.

Both CENAP and CENPES became key to the technical capacitation, improvement, and professionalization of the manpower needed for the oil industry development in Brazil. The objectives of each went on according to the need of the moment and they sought solutions for the new problems caused by the Petrobras growth.

4.4 The National Petroleum Council and Petrobras as Institutions Providing Improvement and Professionalization Courses in Brazil

4.4.1 Petrobras Prepares Its Technical Personnel

Published on 24 September 1957, the Manual with the suggestive name PETROBRÁS *prepares its technical personnel* reports the company acute demand for technicians from different areas, such as chemists, industrial chemists, geologists, engineers of all specialties, economists, management technicians, and skilled workers (Petrobras 1957a). Petrobras, as a company that trained this workforce, tried to demonstrate, through manuals, the safe career that qualified Brazilian technicians could have.

The manuals were elaborated in clear language and were aimed at the interest of the professional, clarifying what were the minimum requirements delimited by the company to join their staff. The selected ones were offered scholarships, transportation to the place of recruitment of the course, as well as the loan of books and materials. Petrobras had as one of the requirements that after the end of the course, the student stayed at least 2 years in the technical board of the company Petrobras (1957a).

In 1957, the Petroleum Geology course was already in operation in Salvador, in collaboration with the University of Bahia, for 2 years, being accepted graduated by the courses of Engineering and by the courses of Natural history. Also, in Salvador, the Petroleum Drilling and Production Engineer course was held with the objective to offer the engineers enrolled in the course a knowledge of specific disciplines and of basic techniques so that, in a short time, they could integrate with proficiency, the work teams of Petrobras (1957a).

In the Rio de Janeiro State, initially, the postgraduate course in Petroleum Refining was held. It lasted for a year and was open to candidates graduated from the higher courses of Engineering, Industrial Chemistry or Chemistry (Fig. 4.2).

The manual, which was informative, also described which courses it was intended to be offered from 1957 by CENAP, being: Oil Equipment Maintenance Engineering and Extension of Improvement (Petrochemical industry, Asphalt, Schist). From 1957, one year after the creation of CENAP, most of the courses were initiated. The intense period of application of them is related to Juscelino Kubitschek Government (1956–1961), which had as characteristic a developmentalist ideology, emphasizing the industrialization of Brazil. It is emphasized that:

> Juscelino Kubitschek has already found the industry formed and a class of industrial entrepreneurs already widely participant in national life. What happened in his government was the acceleration of the economic development, and especially the consolidation of the national industry, by the introduction of sectorial planning, with priorities represented by demands and favors in relation to the basic industrial sectors (Pereira 1963: 20–21).

Fig. 4.2 Cover of the Manual "PETROBRÁS prepares its technical personnel". *Source* Petrobras (1957a)

Another change undertaken by Kubitschek Government was the establishment of economic relations between national and foreign companies, which increased decisively, making it increasingly difficult for Brazilian industrialists to take nationalist positions (Pereira 1963: 23). This was the case of Petrobras, which accepted the participation of foreigners for the development of the technique and for the training of its technicians. In parallel to this, there was still a major obstacle to the presence of foreigners in sectors directly connected to oil.

From now on, we will take some courses of improvement and/or formation (specifically the Geology course) for description and analysis, chosen for their outstanding influence on the creation of courses, both postgraduate courses in universities, and technical training courses in the Brazilian oil industry.

4.4.2 The Oil Refining Course

In the 1920s, countries such as Mexico were showing interest in installing refineries in Brazil. These initiatives were restricted to projects sent to the Federal Government, but which eventually led to the country's own refineries from 1930 onwards, which later led to the acceleration of the process of training specialized manpower in the oil sector.

In 1932, there was the installation of the first oil refinery in Brazil, an initiative of Brazilian, Uruguayan, and Argentinean entrepreneurs. The installation took place in the city of Uruguaiana (Rio Grande do Sul State), using oil imported from Ecuador, Chile and other countries to produce kerosene and diesel. In 1936, another two small refineries were installed: Rio Grande City (Rio Grande do Sul State) and São Caetano do

Sul City (São Paulo State). In 1949, the construction of the Mataripe Refinery (Bahia State), later called the Landulpho Alves-Mataripe Refinery, was begun. In the same year, the construction of the Bernardes Refinery in Cubatão (São Paulo State) was also initiated, inaugurated in 1955, using French equipment.

The refinery facilities not only showed the progress of the oil refining sector in the country but also regional economic changes in the places of installation, with the formation of a working class that was previously engaged in fishing and agriculture, as in the specific case of Bahia State.

In 1950, two technicians from the National Petroleum Council and the Head of the Chemistry Department of the National Faculty of Philosophy (FNFI) idealized the creation of a course to specialize engineers and chemists in oil refining. Thus, they invited Dr. Kenneth Albert Kobe (1905–1958),[18] a chemical engineer and a professor at the University of Texas, to develop the organization project of the course, who suggested the hiring of three foreign professors from the United States: George Fekula,[19] Robert Maples,[20] and Ford Campbell Williams,[21] hired for one year, period when they developed the course program based in Kobe guidance and were then the responsible ones for teaching the classes of what would be the first oil refining course in Brazil (Petrobras 1959b; Caldas 2005).

Thus, in February 1952, it began the first course of Oil Refining in the country, in agreement with the University of Brazil (Caldas 2005). The course period was approximately one year. In the beginning, the theoretical classes were given at the Getúlio Vargas Foundation and the practices carried out in the laboratories of the National School of Chemistry and in the Laboratory of Mineral Production in Rio de Janeiro City. In 1955, in order to ensure the continuity of the course, which was held during 1953 and 1955, the laboratories and other necessary facilities were built on the premises of the University of Brazil, in Rio de Janeiro City. In June of the same year, the task of maintaining and administering the course was transferred from CNP to Petrobras (1959a): 7–9.

During this period, three more refineries started: (1) Manguinhos (Rio de Janeiro State); (2) Capuava (São Paulo State); and (3) Amazon Oil Company (COPAM). It started its operations in 1956, destined to the refining; for that reason, later it was named as Refinery Isaac Sabbá or Refinery of Manaus. The other two started their operations in 1954.

There were also two asphalt and fertilizer plants (both in Cubatão, São Paulo State), as well as the superintendence of Shale industrialization (in Tremembé City, São Paulo State) and the Petrochemical Industry (Rio de Janeiro City—Rio de Janeiro State), among others.

Therefore, the refining courses were directly related to the improvement of manpower along with a sector that was trying to be consolidated in the country: while the oil was not in large quantity, it entered the market through imported oil and its refining.

In the beginning, the classes received a financial aid, as a scholarship to attend the course. However, in 1957, Petrobras modified this picture and began to admit, if they wished, students as employees for a minimum of 1 year after graduation (Petrobras 1959b).[22] The course of 1957 began with a revision course—Introduction to the Petroleum Refining Course—which had as purpose the preparation of technicians and it was "destined for postgraduate students from different school histories". As a result, there was a need to provide students with a previous course, that is, an intensive course of basic subjects, lasting approximately two months (Cenap 1957: 1). Between 1953 and March 1959, seven classes were formed, totalizing 106

[18]Chemical engineer and North American teacher. Consultant for many oil and chemical companies.

[19]No bibliographic references were found.

[20]No bibliographic references were found.

[21]Born on 28 December 1921 in Nanaimo, Canada. Graduated in Chemical Engineering, in mid-fifties he moved to Brazil. He was professor and researcher from CENAP and CENPES. In 1967, left the company, continuing in the chemical research area. He was one of the main responsible for the implementation of the research in Petrobras.

[22]Petrobras (1959b).

students, among them, 95 were hired by Petrobras and the remainder went to private refineries.

The Refining course of 1959 adopted a new policy, being henceforth named as the Refining Improvement Program. It began to be considered a graduation-level course, with the different objective of promoting the improvement of Petrobras own technical personnel, classified by the experience in petroleum refining or petrochemicals. The candidates for the course, belonging to the company, should have more than one year of services provided within an industrial unit of the company and indicated by them (Petrobras 1959b). The first one and a half of the course, considered a further improvement of basic subjects, totaled 24 class hours per week, with the following disciplines (Table 4.3).

The first period of the course lasted 3 months, making a total of 32 h of class per week. It comprised the following disciplines (Table 4.4).

The second period, lasting 3 months and totaling 32 h of class per week, consisted of the following disciplines (Table 4.5).

The third period, lasting 6 months, totaled 44 h of classes per week, and it was composed by the Practical Internship in Refinery (Fig. 4.3).

The process of construction and installation of new refineries continued in the following years: in 1961, the Duque de Caxias Refinery (Rio de Janeiro State) was inaugurated, and the construction work of the Gabriel Passos Refinery (Minas Gerais State) began in 1962. During this period, Brazil also began to receive governors from the US states interested in the technology and operation of the country refineries.

Table 4.3 Disciplines (basic subjects) of the Refining course—1959

Disciplines	Weekly hour/class
Stoichiometry	08 (theoretical)
Materials Technology	03 (theoretical)
Organic Chemistry	03 (theoretical)
English	06 (02 theoretical and 04 speaking)
Thermodynamics	04 (theoretical)

Source Petrobras (1959a: 7–9)

Table 4.4 Disciplines (first period) of the Refining course—1959

Disciplines	Weekly hour/class
Unit Operation of Refinery I	10 (06 theoretical and 04 h/class in laboratory)
Fundamental Calculations of Processing	06 (theoretical)
Fundamentals of Refining I	04 (theoretical)
Equipment of Refining	03 (theoretical)
Tests and Specifications of Petroleum	05 (01 h/class theoretical and 04 h/class in the laboratory);
Technical English I	02 (theoretical)
Notions of Administration I	02 (theoretical)

Source Petrobras (1959a: 7–9)

Table 4.5 Disciplines (second period) of the Refining course—1959

Disciplines	Weekly hour/class
Unit operation of Refinery II	10 (06 theoretical and 04 in laboratory)
Fundamentals of Refining II	04 (theoretical)
Processes of Refining	03 (theoretical)
Instrumentation	05 (02 theoretical and 03 in laboratory)
Technical English II	02 (theoretical)
Thermodynamics of Chemical Engineering	04 (theoretical)
Operations of Refineries	02 (theoretical)
Notions of Administration II	02 (theoretical)

Source Petrobras (1959a: 7–9)

The improvement, through the refining course continued over the years, in specific situations in the refineries, such as the Landulpho Alves Refinery in Mataripe (Bahia State), which in 1963 offered the Refining Course of the Northeast. The beginning of the installation of the refineries in Brazil, at an accelerated pace, especially in the fifties and sixties, in order to serve the market and to improve and professionalize the necessary manpower. In this way,

Fig. 4.3 Cover of the Oil Refinery Course Manual—CENAP 1959. *Source* Petrobras (1959b)

the country was involved in the development of its own petroleum technology, in the construction of its know-how.

This rapid and proper development was only possible because 50% of the net profits were destined to the petroleum research, and these profits were necessarily directed to the research of the national subsoil (Silva 1952); that is, the research took a new turn, turning to the exploration and the prospecting of the soil, a movement opposite to what happened in many countries. It is also worth noting that most of the refineries mentioned are state-owned, such as the Cubatão Refinery, or even with a mixed economy, such as the Mataripe Refinery. However, we also have private refineries in the national territory, but we will not analyze them because the focus of the work is the refineries owned by CNP and Petrobras.

4.4.3 Course of Maintenance of the Petroleum Equipment

In 1958, in agreement with the Institute of Technological Research, the course of Maintenance of Petroleum Equipment was installed in

São José dos Campos (São Paulo). In the first class, a total of 11 technicians were trained. The course was then transferred in 1959 to the Presidente Bernardes Refinery in Cubatão City and later to the Duque de Caxias Refinery in Rio de Janeiro State (Caldas 2005: 29). It aimed to specialize Brazilian engineers in the maintenance of equipment used in the oil industry, developing in them what can be defined as maintenance mentality (Petrobras 1959b). The need for a maintenance engineer was due to the large variety of equipment found in the production fields and refineries. Specialized personnel was needed to keep the entire fleet of equipment in perfect working order, periodically revising or even promoting preventive maintenance, due to the high costs involved in imported equipment (Petrobras 1959b).

In order to apply for the course, it was necessary to have knowledge of mechanics and applied electricity. The registration was made available to mechanical engineers, electricians or even the graduated by the Aeronautical Technological Institute (ITA).

As mentioned, this had a fundamental participation in the development and solutions of the didactic-scientific terms initially faced by Petrobras. ITA who had responsibility for the useful life of oil industry equipment:

> A team of Petrobrás technicians was last month in the cities of Salvador, Maceio, Recife, Campina Grande, Fortaleza, Belo Horizonte and Porto Alegre, recruiting engineers of any specialty (except agronomists) to the Landulpho Alves Refinery board, where they will undergo a period of improvement, specializing in maintenance of equipment (Recrutamento de … 1961: 26).

The Petroleum Equipment Maintenance course (Petrobras 1959b) lasted 8 months, consisting of an introductory period of 2 months, which contained the following subjects (Table 4.6).

After this period, the period named Specialization started, with a duration of 3 months, consisting of the following subjects (Table 4.7).

The second period of the Specialization, also lasting 3 months, was composed of the following subjects (Table 4.8).

Table 4.6 Disciplines (introductory course) of the Petroleum Equipment Maintenance course—1959

Disciplines	Weekly hour/class
Applied Electrotechnical	04 (theoretical)
Thermodynamics	04 (theoretical)
Unit Operation	04 (theoretical)
English I	05 (theoretical)

Source Petrobras 1959b

Table 4.7 Disciplines (first period) of the Petroleum Equipment Maintenance course—1959

Disciplines	Weekly hour/class
Lubrication	30 h (1 week isolated)
Refining Equipment I	05 (theoretical)
Maintenance Organization I	04 (theoretical)
Equipment Inspection	02 (theoretical)
English II	03 (theoretical)

Source Petrobras (1959b)

Table 4.8 Disciplines (second period) of the Petroleum Equipment Maintenance course—1959

Disciplines	Weekly hour/class
Refining Equipment II	03 (theoretical)
Maintenance Organization II	03 (theoretical)
Standards and Specifications	03 (theoretical)
Diesel Engines Maintenance	80 (2 weeks)
Welding	60 (2 weeks)
English III	03 (theoretical)

Source Petrobras (1959b)

At the end of the course with approval, the students received a certificate of proficiency, which allowed them to be assigned to work in one of Petrobras industrial units (Petrobras 1959b). "In 1962, the Northeast Maintenance Course (CMN) and the Special Maintenance Course (CEM) were started, both in Bahia State" (Petrobras 1959: 29) (Fig. 4.4).

4.4.4 Course of Oil Engineering

In 1963, the first course of Petroleum Engineering in Salvador (Bahia State) was held, in

Fig. 4.4 Cover of the manual "Petroleum Equipment Maintenance course"—CENAP, 1959. *Source* Petrobras (1959b)

collaboration with the University of Bahia. The course, formerly known as the Petroleum Drilling and Production Course, was designed to improve (at postgraduate level) higher level Brazilian technicians in different sectors of the oil industry. One of the relevant aspects of this course was the lessons of petroleum mining techniques, crucial in the exploration and consequent expansion of the producing fields (Petrobras 1963).

One of the main needs and roles of the oil engineer, at that time, who lacked professionals in this area, was concerned with extracting "the fluid (gas, oil) from the rocks that enclose it and end with its arrival to the refinery", "and he must carry out the series of operations in the most rational and economic way" (Petrobras 1963: 7–8). Thus, "technicians were prepared for the following types of activities: (a) drilling; (b) production; (c) storage and transport; (d) reservoir engineer" (Petrobras 1963: 8).

The characteristics of the course teaching are similar to those of the other courses offered by Petrobras, always trying to fulfill a full-time regime and presenting an up-to-date curriculum (due to the constant development of the oil industry and related fields). An example of this was controlled and compulsory attendance in theoretical and practical classes, as well as student assessment through tests and assignments.

The difference, in relation to university courses itself, lays in the use of "problems derived from the industrial reality as a working instrument and a means of forming in the student the habit of facing and solving new problems safely" (Petrobras 1963: 10).

The duration of the course was approximately one and a half year, comprising six distinct periods: I—Introductory period (8 weeks); II—First Period (12 weeks); III—Second Period (12 weeks); IV—Third Period (12 weeks); V—Fourth Period (8 weeks); and VI—Fifth Period (16 weeks). The disciplines offered in these six periods were: Introduction to Petroleum Engineering; Physico-chemical; Geology I, II, III, IV; Reservoirs I, II, III; Drilling I, II, III; Production I, II; Evaluation of Formations; Projects, Operations and Programs; and Technical English I, II, III and IV. Field training activities were generally carried out in the Region of Bahia for a period of 16 weeks.

The first subject of the course—Introduction to Petroleum Engineering—draws attention, with themes related to the following subjects: geography and history of petroleum; Law No. 2.004 (creation of Petrobras); stages of industry; structure of Petrobras; roles of the petroleum engineer; and analysis of the last report of the activities of Petrobras. This was one of the only courses known to have offered the discipline of introduction to petroleum engineering, mentioning and demonstrating the relevance of knowing the structure, history, and territory in which one works (Fig. 4.5).

The maximum age to apply for the Petroleum Engineering course was 39 years. In addition, "due to a series of factors inherent to the activities of a Petroleum Engineer, and based on experiences of the Company, it is not allowed the entry of females in the Course" (Petrobras 1963: 21). This was repeated in the Petrobras manuals, in which it was pointed out that the conditions in which an oil engineer lives, would not be propitious to the female gender:

> the stage in which the minor are the conveniences is that of the preliminary intern in the field, when the graduate will work, as a skilled worker, in the drills and in the productions activities and pipelines. It is a rather rough period, and the work can be day or night, outdoors, with great energy expenditure, since physical work is intense. That is why Petroleum Engineers should be highly disposed and men with good health conditions (Petrobras 1963: 21–22).

It was also pointed out that work was carried out far from the great centers, without many resources or the structure offered by the urban regions. The insertion of women in the labor market was therefore barred by numerous justifications, which did not facilitate their entry into higher education courses.[23]

Petrobras manual related to the Petroleum Engineering course presents a speech that claims that the company was for everyone, that the benefits achieved would be "our benefits", "profits and progress for the whole nation", and that the work developed Petrobras employees would represent "an important part of the consolidation of the country economic development" and "improved of the conditions of our people life" (Petrobras 1963: 24). This was a way to convince new engineers to take part in Petrobras training courses, becoming employees of Petrobras, pointing out career benefits and appealing to patriotic sentiments.

As mentioned in the first and second parts of this book, the development of the oil industry in the nationalist context is very similar in Mexico and Brazil. However, in the process of improvement and professionalization, the difference is clear, as we intend to demonstrate in relation to petroleum engineering, in the creation

[23]It should be noted that, at that time, there was no technology and machinery that aimed to "soften" the worker's physical effort. The tools of work were still rudimentary, with the objective not of worker safety and health, but of increasing productivity and, consequently, profits. That is why, when the Consolidation of the Labor Laws (CLT) was approved on 1 May 1943, the objective was to protect women, reserving to it an exclusive chapter entitled "Protection of Women's Work". Among the various provisions elaborated in this chapter, the employer was barred from employing the woman in service who demanded the use of muscular force greater than 20 kg for continuous work or 25 kg for occasional work. In other passages, the legislator also made clear the intention to protect the work of women, as occurred in work in underground mines, provided in CLT article 301 that it was only allowed to men aged between 21 and 50 years.

Fig. 4.5 Cover of the Manual of Petroleum Engineering course—CENAP, 1963. *Source* Petrobras (1963)

of the courses and the way in which it has consolidated in both countries.

In this case, comparing the description of Brazil with Mexico, the first oil-related careers were the professions of chemist and petroleum engineer, both consolidated in the Faculties of Chemistry and Engineering in 1927 (Crisolis Reyes 2011). Both professional branches worked together to prepare the study plans for the first courses of the professions mentioned above, which resulted in the following curriculum design (Table 4.9).

The oil-related professions (chemists and engineers) in Mexico were encouraged from a post-revolution government nation project (1917). In that period, the country was already an oil producer, and investments in the area were essential in Brazil the difference is clear: it has been sought for oil since the beginning of the nineteenth century; subsequently, refineries are created; Finally, the policy of professionalization and improvement is exercised through the creation of the CNP and Petrobras, two scientific

Table 4.9 Disciplines of the Petroleum Engineering course, Mexico

First year	Mathematics; Topography and Practice; Descriptive Geometry; Static; First course of Stability; Topographic Design; Architectural Design
Second year	Dynamics and Mechanisms; Notions of Mineralogy and Geology; Hydraulics; Thermal Machines; Oil Exploration; Practice; First Course of Construction Procedures, Design of Machines
Third year	Electricity; Concrete; Sanitary Engineering; Roadways and Waterways; Practice; Contracts and Legislation; Exploitation of Oil
Fourth year	Oilfields: General Practice

Source Crisolis Reyes (2011: 89)

and industrial institutions that begin the qualification process of the petroleum sector.

4.4.5 Course of Introduction to Geology and Petroleum Geology

After the creation of the Petroleum Engineering course, it began the phase of training geologists, a basic need to continue studying and exploring the Brazilian territory in search for oil (Mattoso 2012). In order to continue with the training plan for specialized technicians, CENAP organized and held in Salvador, Bahia, during the period from 1 April to 30 June 1957, a program of studies for the better preparation of the approved students for the first course of introduction to geology and that would be destined to the course of Petroleum Geology, to begin in July of the same year (Cenap 1957b: 1). It should be noted that the Exploration Department had an active participation with the CENAP to prepare the course (Fig. 4.6).

The program of the course of introduction to Geology had as main subjects English, Mineralogy and Petrography, Topography and General Geology, which were complemented by lectures, seminars, guided reading, and arguing about the geology of Brazil. The conferences and seminars covered the following themes: General Geology of Brazil; Brazilian Crystal; Parana Basin; Amazon Basin; Mid-North Basin; and Heavy Minerals for Geological Correlation (Cenap 1957b: 2).

For this course, there were 20 vacancies, which were divided into two classes for better use, with each class lasting 2 hours. Brazilian and foreign teachers were also hired to teach the various disciplines (Cenap 1957b: 2), as it was the case of the one that addressed the Parana Basin, to which the Brazilian paleontologist Frederico Waldemar Lange was invited (1911–1988),[24] who gave a series of lectures on the geology of the Parana Basin. "Prof. Lange was also kind enough to offer to the Course a collection of 44 samples of rocks typical of the Parana Basin, as well as several leaflets and publications of interest for the Course in question" (Moggi 1957: 1). As there was the presence of Americans among the teachers and part of the material was in English, the English discipline was mandatory—the only one, in fact, taught every day, including Saturdays (Leal 2008: 253).

In July 1957, the first course in Geology was organized sequentially to the course of introduction to Geology, with an emphasis on the study of Petroleum, a pioneer initiative in the formation of geologists in Brazil. The objective of this course was to form and train the geologists that Petrobras so badly needed. Professor Irajá Damiani Pinto, graduated in Natural History from the University of Rio Grande do Sul (URGS, now UFRGS), was invited to organize the course of Petroleum Geology and to participate in the initial phase of the project—the selection of teachers who would compose it (Pinto 2011).

The Geology course lasted for 2 years (in an integral and dedicated regime) and included "all

[24]Born in Ponta Grossa, Paraná State, Lange graduated in Economics and Accounting by the Higher Institute of Commerce of Curitiba and specialized in Paleontology at the Paranaense Museum, where he was appointed director. In 1955, he joined Petrobras as a surface geologist, reaching the top position in the Exploration Department in 1960. He pioneered Micropaleontology in Brazil. Published 22 articles.

ESQUEMA PROVISÓRIO PARA O PERÍODO DE 1 DE ABRIL A 30 DE JUNHO

1ª sem.	2ª e 3ª sem.	4ª sem.	5ª e 6ª sem	7ª sem	8ª e 9ª sem.	10ª sem.	11ª e 12ª sem.	13ª sem
Inglês ARCABOUÇO GERAL DA GEOLOGIA DO BRASIL	Inglês Mineralogia Petrografia Topografia Geol.Geral	Inglês CRISTALINO BRASILEIRO	Idem 2ª e 3ª semanas	Inglês BACIA DO PARANÁ	Idem 2ª e 3ª semanas	Inglês BACIA AMAZÔNICA	Idem 2ª e 3ª semanas	Inglês BACIA DO MEIO NORTE

1ª, 4ª, 7ª, 10ª e 13ª semanas (5 semanas):

Inglês – Todos ós dias
Seg.,Ter.e Quar.– Seminários
Quintas – Leitura orientada sôbre o assunto dos Seminários.
Sextas – Arguição e notas sôbre o assunto

ATIVIDADES DURANTE AS SEMANAS ACIMA

Nas demais semanas (8 semanas):

SEGUNDA	TERÇA	QUARTA	QUINTA	SEXTA
A Inglês B Min. e Petr. A Topografia B Geol.Geral	B Inglês A Min. e Petr. B Topografia A Geol.Geral	A Inglês B Min.ePetr. A Topografia B Geol.Geral	B. Inglês A Min. e Petr. B Topografia A.Geol.Geral	Excursão Arred. de Salvador Demonst. Relatór.

Fig. 4.6 Provisory program of the Introduction to Geology course. *Source* Petrobras (1957b)

Table 4.10 Disciplines of the first year of the Petroleum Geology course

First year	Duration disciplines
Introductory period	8 weeks descriptive geometry; Physical and Geological Sciences; General Chemistry; and English
First period	12 weeks Mineralogy; Physical Geology; Historical Geology; and English Technical
Second period	12 weeks Petrology; Geomorphology; Paleontology; and Technical English
Third period	12 weeks Optical Mineralogy and Petrography; Structural geology; Stratigraphy; and Technical English

Source Perguntas e … 1959: 11

the geology subjects required to obtain the B. S. (Bachelor of Science) at several major universities in the United States" (Humphrey 1961: 1). The course received candidates from the following areas: civil engineering and mines, chemistry and agronomy. The majority of the participants came to work in Petrobras itself or even in Brazilian universities (Sial 2008) (Tables 4.10 and 4.11).

Table 4.11 Disciplines of the second period of the Petroleum Geology course

Second year	Duration (weeks)	Disciplines
Practical period	8	Field Geology
First period	12	Sedimentary Petrography; Interpretation of Maps and Photogeology, Oil Geology; and Technical English
Second period	12	Upper Mineralogy and Opaque Minerals; Paleogeology; Economic Geology; and Technical English and Geology Problems
Third period	3	Subsurface Geology; Sedimentation; Exploration Geology; Technical English and Geology Problems

Source Perguntas e … (1959: 11)

As mentioned, to teach the disciplines, Brazilian and foreign teachers were hired. Brazilian teachers were selected by interview and, in some cases, the hired ones assisted foreign teachers. It is possible to mention among them:

(1) Murilo Cabral Porto, from the Faculty of Philosophy/USP, to assist Max Carman. The Faculty of Philosophy prepared staff to take on the position of professors, whether in the university or in the courses of the "elementary school" and "scientific" (as the courses for students from 11 to 18 years old were called at the time in Brazil, so after the primary school with students from 6 to 11 years), in the state of São Paulo. Murilo works in an exploitation and drilling company for water wells.
(2) Salustiano Oliveira Silva, a mining and civil engineer, graduated from the School of Mines of Ouro Preto/Minas Gerais State, a geologist who had worked for the National Petroleum Council (CNP) succeeded by Petrobras, and had extensive experience in geological surveying in Maranhão and, especially in the Amazon State, assisted in the teaching of Petroleum Geology and Structural Geology.
(3) Sylvio de Queirós Mattoso, a mining and metallurgist engineer from the Polytechnic School USP, who worked in mineral exploration for the ceramic industry in the private sector, for which he worked from the Rio Grande do Sul State (under clays) to Ceará State (in the research of magnesite and chromite) (Mattoso 2012: 1).

The three joined Petrobras in 1957 and 1958. In 1959 or 1960, "mining engineer and metallurgist, Shiguemi Fujimori, also from the Polytechnic School of University of São Paulo, newly formed or formed a year before, was hired as an auxiliary" (Mattoso 2012: 1).

The foreign professors that were hired came mainly from the University of California (UCLA), the same university where Fred la Salle Humphrey came from, who directed the course of Petroleum Geology until its closure (1964) and indicated the other professors to be hired by CENAP (Mattoso 2012):

> The professor of Petroleum Geology was Cordell Durrel, I believe also from UCLA, in charge of Structural Geology, Petroleum Geology and another one that escapes me from memory at the moment. Donald Briant came from Tucson, Arizona (University of Arizona), in charge of ministering the courses of Stratigraphy, Paleontology and Historical Geology; Max Carman, I believe that also came from UCLA that was with Mineralogy, Petrography (Mattoso 2012: 1).

The internship (Field Geology) generally occurred in the State of Sergipe, in the region that encompasses the municipalities of Itabaiana until Laranjeira, near the capital Aracaju (Sergipe State), including crystalline and sedimentary basement rocks. Humphrey, the course coordinator, placed maximum emphasis on fieldwork. All aspects of geology that could be related to the oil deposits were emphasized, and many exercises were carried out (Mattoso 2012).

The course was maintained by CENAP and was carried out in Salvador in agreement with the University of Bahia, in continuation to the course of introduction to Geology, as we have already said. the divulgation of the courses, as well as the diplomas of the technicians by Petrobras, were subjects always present in the technical bulletins of Petrobras and newspapers of the daily press.

CENAP Geology course was closed in 1964 by a number of factors, such as the anti-American wave of the period—somewhat blocked by the beginning of the military regime in 1964—and almost at the same time in 1957 by the foundation of the first Geology courses in the country, which were already forming a proper number of geologists in Brazil for the period (Mattoso 2012) (Fig. 4.7).

NOVA TURMA DE geólogos de petróleo

MARCADA A DIPLOMAÇÃO PARA O DIA 3 DO CORRENTE

Está marcada para o dia 3 do corrente, em Salvador, a cerimônia da diplomação de mais uma turma — a segunda — do Curso de Geologia de Petróleo, mantido pela PETROBRÁS, naquela capital, em convênio com a Universidade da Bahia. A nova turma se compõe de 14 elementos.

Fig. 4.7 Informative regarding the graduation of the Petroleum Geology course—December 1959. *Source* Interesse dos … (1959: 7)

4.5 The Influence of the Petrobras Geology Course in the Geologist Training Campaign by the Federal Government in 1957

The organization of the course of Geology by Petrobras was one of the main bases for the Brazilian Universities to organize regular courses in the area, with four years of duration (Humphrey 1961):

> In December 1960, the Universities of São Paulo and Porto Alegre formed their first group of geologists. It was possible to train this first group, only three years after the beginning of the course, in view of their students already possessing the knowledge base corresponding to the first year of work, acquired in the courses of Natural History, Sciences or Engineering, in which they were previously enrolled (Humphrey 1961: 1).

In the same year, Petrobras modified the duration of the Petroleum Geology course, which was 2 years, transforming it into a 12-month postgraduate course, the graduates of the 1960 course being the first geologists graduated by the Universities of São Paulo, Porto Alegre and Ouro Preto (Moggi 1961).

The curriculum of the universities generally did not yet have a complete geology training program such as the one contained in the Petrobras Geology course. In order to overcome this deficiency, behind the creation of geology courses in universities, the Federal Government instituted a Campaign for Geologists Formation (CAGE) in 1957. CAGE temporarily supplemented the programs of these universities, helping them to obtain foreign teachers to supplement the currently limited number of Brazilian Geology teachers (Humphrey 1961: 2):

> In 1961, Petrobras considered transferring the postgraduate courses of CENAP to the Universities. But as the newly formed geologists entered Petrobras, their deficiency was perceived in Petroleum Geology, Paleogeology, Sub-surface Geology, Sedimentary Petrology, Sedimentation, Principles of Geophysics, Field Geology, Photogeology, and, in some cases, Structural Geology (Humphrey 1961: 2).

Soon, Petrobras Geology course continued with its postgraduate course for the improvement and training of graduates in Geology by the universities. In 1961, the didactic coordinator of CENAP courses in Bahia and the Course of Geology, F. L. Humphrey, emphasized that one of the major shortcomings of undergraduate in Geology of the universities was in Field Geology, as there was no actual experience of field geological mapping (Humphrey 1961: 3). For Humphrey, it was indispensable that the Petrobras geologists had a good training in Field Geology, mentioning as an example the case of the United States, where few geologists hired by companies did not have a full postgraduate degree, requiring at least a Master of Science degree in Geology (Humphrey 1961: 3):

> CENAP would therefore have to recruit future postgraduates of Geology, among people of little tradition in the study of this branch of Science. A whole didactic structure was set up, already to provide the national frameworks that should, in the short term, replace the foreign contracted to the tasks of the routine of the prospection already to arise the vocations scientifically oriented to the research. Still using an alien faculty member, CENAP was able to prepare until 1961, 66 geologists (Oliveira 1961: 143).

To conclude, Humphrey also stated that it was not possible to rely on Brazilian universities for the acquisition of properly trained geologists. The idea of implanting Geology courses in Brazil began in the fifties. Some universities, in isolated attitudes, began to create commissions or projects for the creation of undergraduate courses of Geology:

> In 1955, the University of Rio Grande do Sul (URGS) created a commission to study the creation project of a Center for Geological Studies and Research. In the same year, the University of São Paulo (USP) elaborates a project to create a Geology course to be assessed by the legislature of that State (Azevedo and Terra 2008: 375).

"Discussions around the theme gained strength in late 1956, with the Ministry of Education and Culture designating a commission to evaluate the creation of the first courses of

Geology in Brazilian universities" (Azevedo and Terra 2008: 375):

> A preparatory committee was set up, which, after conducting a research on the labor market, chose to indicate as a solution the creation of specific Geology courses, which should form, within the next few years, around 700 (seven hundred) geologists, in order to supply a growing market (Entrevista Viktor … 1982: 57).

In December 1956, "the Coordination of improvement of personnel of higher level (CAPES) takes to the minister Clóvis Salgado the proposal of creation of four courses of Geology in the Country, headquarter in Ouro Preto, São Paulo, Recife and Porto Alegre cities" (Azevedo and Terra 2008: 375).

On 18 January 1957, the current President of the Republic, Juscelino Kubitschek de Oliveira, instituted a Campaign for Formation of Geologists in the Ministry of Education and Culture, through Decree No. 40.783, which aimed to ensure the existence of personnel specialized in Geology in the public and private enterprises, in quality and sufficient quantity to the national necessities (Brasil 1957).

The activities of the CAGE would be financed with funds from a special fund managed by the Ministry of Education and Culture and constituted of: allocated contributions in the budgets of the Union, states, municipalities, parastatal entities and mixed capital companies (such as Petrobras); contributions from contracts and agreements with public and private entities; and donations, contributions, and legacies of individuals (Brasil 1957).

CAGE provided material resources, "studied and planned the budgetary needs for the acquisition of didactic material and laboratories, equipment for fieldwork, including vehicles" and for the hiring of teachers, initially to four higher education institutions (Gomes 2007: 61). The Campaign also supported the need to bring teachers of various specialties to Brazil. according to the Brazilian engineer and geologist Othon Henry Leonardos,

> When CAGE founded its courses, there was no Brazilian element capable of giving a course in structural geology. We went to get a teacher at University of Tehran, his name is Prof. Louis de Loczy, former director of the Geological Service of Prussia, who has worked in every country in the world and who has the luxury of being Brazilian, Brazilian nationalized. So, he was in Persia, but he was Brazilian. (Encontro de Geólogos 1966: 86).

The payment for foreign teachers was made in US dollars, which is expensive for the CAGE budget, but necessary. In addition, it was difficult to hire foreign teachers, especially in the area of Geophysics.

Generally, admitted students were entitled to the scholarship[25] provided by CAGE. They should devote themselves to a full regime, 8 h per day, and could not be disapproved, with a minimum score of 6.0. The internship of these students was guaranteed by Petrobras or other government agencies (Fig. 4.8).

The first geology courses in Brazil were created at the University of Ouro Preto, at the University of Rio Grande do Sul, at the University of Pernambuco and at the University of São Paulo. The first classes of geologists were formed in Porto Alegre, in São Paulo and in Ouro Preto cities in 1960 (Fig. 4.9).

Petrobras was cordially involved with CAGE, through support and agreement signed in 1957 for the creation of geology courses in Brazilian universities. After the complete transfer of training courses to geologists, the Ministry of Education and Culture promoted the extinction of CAGE in 1965. In the early sixties, after the formation of the first geologists, Petrobras absorbed 50% of professionals formed by universities (Lange 1962).

Even with the transfer of the training of geologists to the universities, Petrobras continued its role since the fifties, now carried out through accredited representatives of the Production department, who went to the Geology Schools and were in charge of interviews with all the graduates who had an interest in joining them

[25]Vitor Leinz stated, in the First Meeting of Geologists in 1966, that CAGE, in São Paulo City "for two or three years provided scholarships in the form of loans. The students signed and committed to return it." (Encontro de geólogos 1966: 118).

Fig. 4.8 Bus of the CAGE Campaign—University of São Paulo (The bus financed by CAGE arrived at the University of São Paulo in the second semester of 1958). *Source* Ônibus financiado … 1959

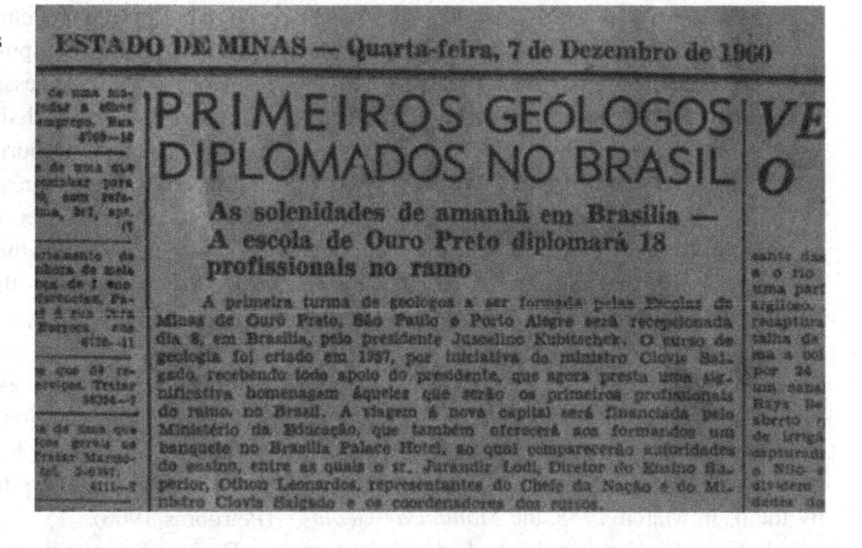

ESTADO DE MINAS — Quarta-feira, 7 de Dezembro de 1960

PRIMEIROS GEÓLOGOS DIPLOMADOS NO BRASIL

As solenidades de amanhã em Brasília — A escola de Ouro Preto diplomará 18 profissionais no ramo

A primeira turma de geólogos a ser formada pelas Escolas de Minas de Ouro Preto, São Paulo e Porto Alegre será recepcionada dia 8, em Brasília, pelo presidente Juscelino Kubitschek. O curso de geologia foi criado em 1957, por iniciativa do ministro Clovis Salgado, recebendo todo apoio do presidente, que agora presta uma significativa homenagem àqueles que serão os primeiros profissionais do ramo no Brasil. A viagem à nova capital será financiada pelo Ministério da Educação, que também oferecerá aos formandos um banquete no Brasília Palace Hotel, ao qual comparecerão autoridades do ensino, entre as quais o sr. Jurandir Lodi, Diretor do Ensino Superior, Othon Leonardos, representantes do Chefe da Nação e do Ministro Clovis Salgado e os coordenadores dos cursos.

Fig. 4.9 First geologist graduated in Brazil—Minas Gerais State Newspaper, 1960. *Source* Primeiros geólogos … (1960)

in geological and geophysical activities. The interviews were based on the candidate's curriculum, with no proof for selection. The geologist started as a trainee technician and, after two years of field practice, could have the opportunity to be assigned to courses or internships in the country or, possibly, abroad (Esclarecimentos aos … 1965).

It is worth noting that in 1947 the First Brazilian Congress of Geology in Rio de Janeiro had been carried out in this process of formation of the Geology course, even before its creation by Petrobras, and that it continued every year until 1974, then taking place every 2 years.

In March 1966, what we can define as a new phase for the geology courses was held in Porto

Alegre City: the I Meeting of Geologists, with a board dedicated to the situation of geology teaching in Brazil, coordinated by the Professor Josué Camargo Mendes.[26] The great point of discussion at the board was the postgraduate courses and the need, of foreign experts to teach disciplines little familiar to Brazilian geologists (Encontro de geólogos … 1966). Professor Josué Mendes stated that "there is no possibility of people living in science without contact with foreign universities. This need will remain" (Encontro de geólogos … 1966: 83). The proposal of the event was to maintain the link with foreign institutions for a greater exchange of visiting professors, internships and other means of communication.

These Congresses and Meetings accelerated the process of consolidation of the natural sciences in the country, contributing to the training of professionals and intellectuals in this area, having the opportunity to raise major problems such as teaching and covenant with foreign universities.

4.6 In Addition to the Improvement and Professionalization Courses

4.6.1 Manual of Surface Geology

In addition to the creation of courses, Petrobras also invested in the preparation, or even in the adapted translations, of its own study material, and in 1958 started a program of specialized separate publications.

In the second part of this book, we refer to the Exploration Department (DEPEX). Here we will try to describe an example of the material made by them. In March 1958, the *Manual de Geologia de Superf'icie* was published, which was an adapted developed updated translation of the Surface Geology Manual, published and elaborated also by the Department of Exploration. The Manual was elaborated on the basis of intensive team work and relied on the collaboration of all the regional exploration districts of Petrobras in order to form its know-how. It was reissued in 1966 by the Exploration Division Surface Sector (DIVEX), an organ of Department of Exploration and Production (DEXPRO), formerly DEPEX. Such publication, this time, occurs by the CENPES.

The Manual aimed to standardize basic procedures in the field of surface geology. The different phases of the field mapping, the description and presentation of data, the equipment and the geological illustrations were treated in order to offer the surface geologist a field guide with useful suggestions to the work, since the study of the area could be applied to the petroleum research, being one of the basic elements of the discovery of sedimentary basins, for that reason it had such importance (Petrobras 1967a) (Fig. 4.10).

The surface geology teams of this period were composed of a chief geologist, an assistant geologist, a trainee geologist (non-obligatory), a surveyor (non-obligatory), two surveyor aids (when there was a surveyor), a driver-mechanic, a cook and a helper (Petrobras 1967a). To gather the geological equipment needed for the research, it was necessary to know: type of mapping; geographical situation; means of communication and transport; and existing facilities in the programmed area (roads, distance from industrial centers, cities, etc.) (Petrobras 1967a).

The Manual established a standard to be followed, demonstrating the need to have solid geological knowledge, especially in relation to the stratigraphic and structural aspects of the sedimentary basins in exploration, being indispensable that, in addition to the surface geology, geologists would have knowledge of geophysics and subsurface geology for exploration activities (Petrobras 1966).

During the years 1957 and 1967, surface geology could be identified as a phase of the evolution of mapping methods, with the use of aerial photography contributing to the discovery of new exploration areas by Petrobras 1967.

Other manuals have been published, such as the one from Gerhard Ludwig[27] entitled New

[26]Professor of Paleontology at the University of São Paulo—USP.

[27]Geologist.

Fig. 4.10 Cover of the Manual of Surface Geology (1966). *Source* Petrobras (1966)

Stratigraphic Division and Faciological Correlation through small internal structures of the Silurian and Devonian sediments in the 1964 Middle Amazon Basin. Other investments were made along the same lines in technical monographs related to science, technology and petroleum, which included: oil exploration; oil drilling (drilling, production, reservoirs); oil refining; petrochemical; terminals and transportation of oil; marketing and distribution; oil equipment; and technological research (Ludwig 1964).

4.6.2 The Work in Bahia State and the Accomplishment of the 1st Meeting of Technical Studies of Petroleum

In 1954, Petrobras carried out its activities in Bahia State through the Bahia Production Region, the Landulpho Alves Refinery in Mataripe, the expansion works of the Landulpho Alves Refinery, CENAP and National Tanker

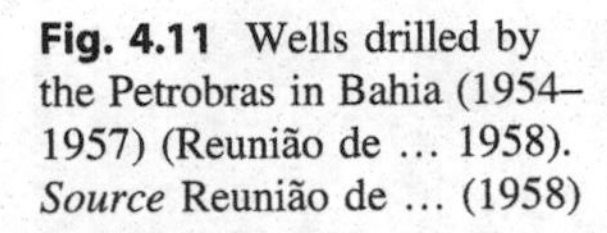

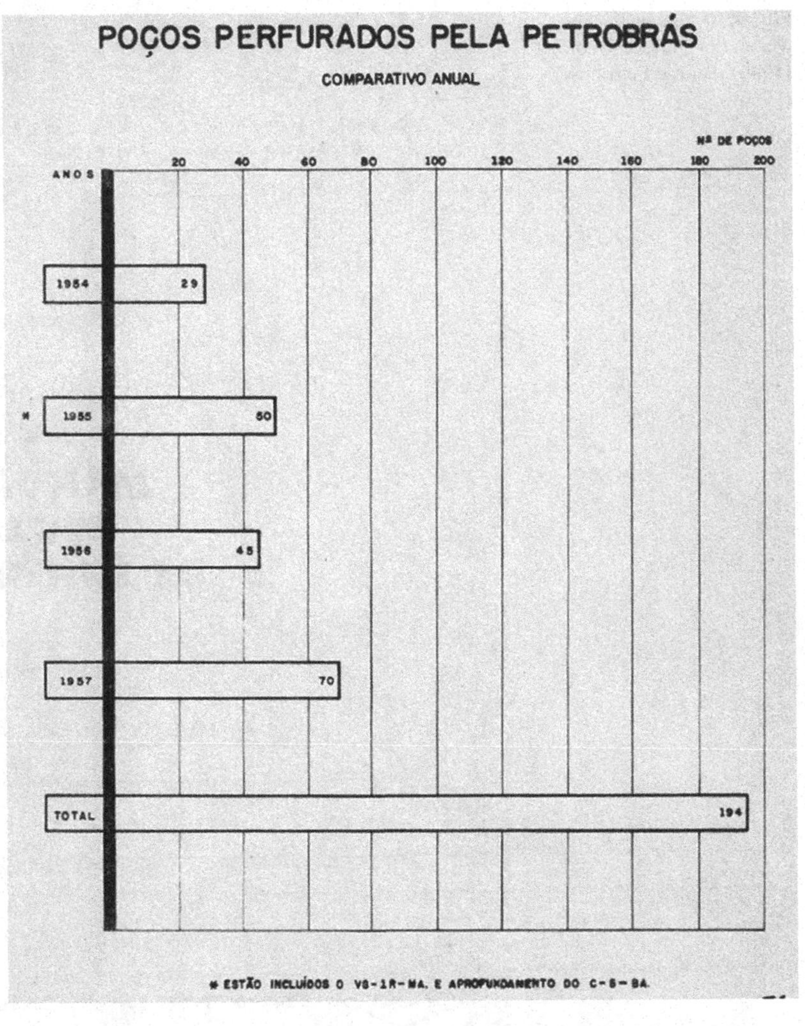

Fig. 4.11 Wells drilled by the Petrobras in Bahia (1954–1957) (Reunião de … 1958). *Source* Reunião de … (1958)

Fleet (FRONAPE).[28] In the second half of 1958, Bahia State exploration department had only three teams of surface geology, four of geophysics (seismic method) and one for small stratigraphic holes (Reunião de … 1958). Most of the materials and equipment used in drilling, in the laboratories and in the structure of Petrobras itself, and in the expansion of the Bahia Refinery were imported from the United States, France, Germany, and the Netherlands.

For a better understanding of the pace of work carried out in Bahia, we will analyze two figures (Fig. 4.11).

[28]In 1958, FRONAPE maintained an office in Salvador, an active operation. On average, 12 domestic and foreign oil tankers arrived monthly.

In this, we can observe a considerable increase in drilled wells in 1957 and state that this increase coincides with the formation (improvement and professionalization) of classes until 1957, including courses in Refining, Equipment Maintenance and Drilling (the course of Drilling became Petroleum Engineering in 1959). In 1958, these numbers continued to grow in the part of drillings: 25 drills were in operation, and 12 pioneer and stratigraphic wells were drilled in search of oil (Reunião de … 1958) (Fig. 4.12).

At the same time, as this figure only confirms the advances made between 1956 and 1957, both in the continuation of drilled wells and in the new drilling of pioneer wells, we perceive the clear

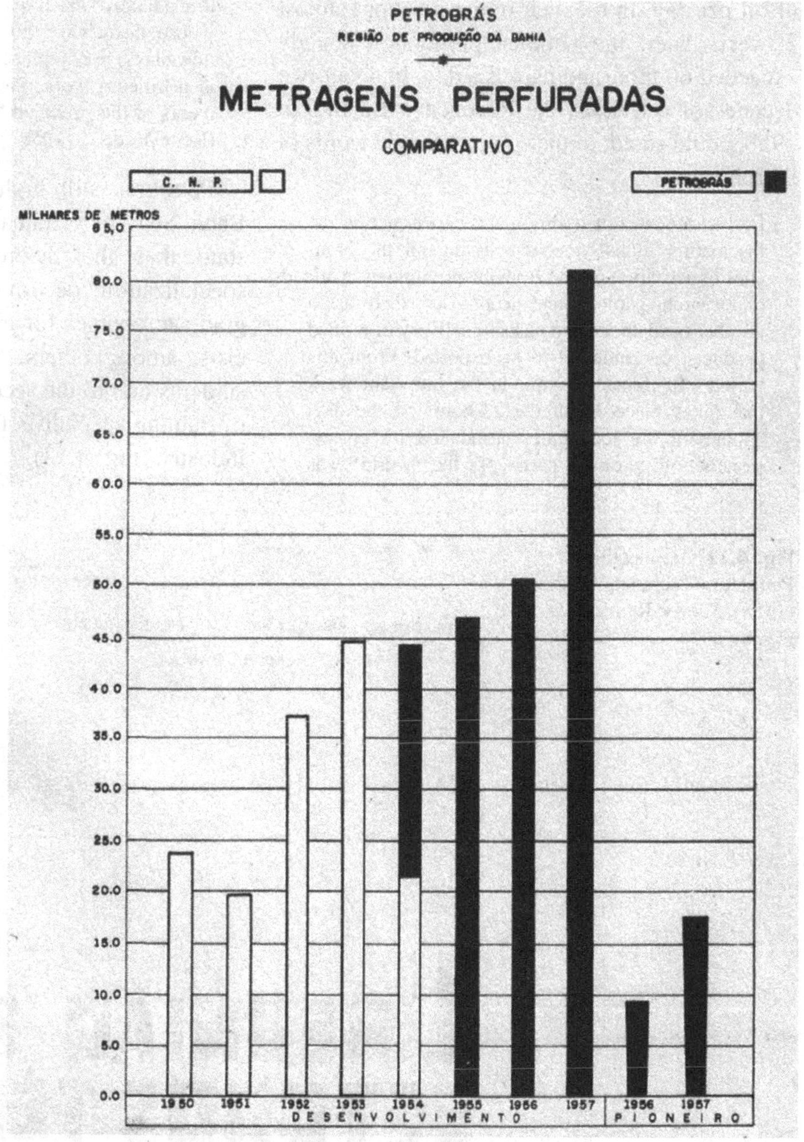

Fig. 4.12 Footages drilled in Bahia State (1950–1957). *Source* Reunião de … (1958)

contribution of the work carried out by CNP. As in the second part of this book, we observed a growth in oil exploration, exploitation, and the oil industry from the second half of 1950, through political changes, investments, and also the return of refineries and new commercially exploitable wells in Brazil.

In 1958, it was also invested in the first meeting of Petroleum Technical Studies (Reunião de … 1958). The event, which took place between October 5 and 11, could not be anywhere but in the first oil-producing province of Brazil (Reunião de … 1958), where the majority of the courses of improvement and professionalization were concentrated in the period.

On 19 January 1939, the well located in the district of Lobato recorded a yield of three barrels of oil per day. In the year of the meeting, almost 20 years later, the national production already exceeded 60 thousand barrels a day. In about two decades of activities, the Recôncavo Baiano, in 1958, could stand, in the somewhat epic words of Petrobras:

> […] as a center of tradition and experience in the fascinating industry, encompassing all the branches of activity it offers: from the preliminary work of research, geology and geophysics, there are a number of methods and tasks to delivery of refined products, or crude oil to be exported. From this varied experience, it gains in the incessant work that takes place in all the 24 hours of the day, enhancing the technique, introduced by professionals of various parts of the world was immediately seized and executed by the nationals whose hosts were growing, to the point where, nowadays, in any place of the country where there is petroleum work, there are always present elements of the so-called "School of the Recôncavo" (Reunião de … 1958, p. 3).

Petrobras still highlights the importance of Bahia State in technical terms, since it was in the state that the theoretical-practical petroleum specialization occurred and evolved through graduate courses for engineers, chemists, physicists, among others, as well as for university students and to the secondary grade, in the sense of training specialists for the various branches of industry (Fig. 4.13).

Fig. 4.13 1st meeting of Petroleum Technical Studies (1958). *Source* Reunião de … (1958)

The purpose of the First Meeting was to publicize the development of research already conducted in the country, providing and receiving suggestions for the progress of the oil industry and, of course, Petrobras. The main topics discussed at the meeting were: oil geology; geophysics; geochemistry; drilling; reservoir and production; collector system, transport and distribution; chemistry and refining; petrochemical; industrial management; economy and organization; medicine; and law and legislation. The main participants of the event were the holders, that is, all the technicians who worked in the oil industry, Brazilian and foreign, special guests and students. The difficulties faced were also dealt with, especially in the case of transportation of equipment and other materials needed to start the exploration work.

From the emergence of Petrobras to its establishment as an oil producer, the advances in the fifties were significant and guided the exploration and consolidation of a national oil industry.

4.6.3 Exploration of Oil in Bahia State in the Early Sixties[29]

See Figs. 4.14, 4.15, 4.16, 4.17, 4.18, and 4.19.

4.6.4 The Courses Continue Through Petrobras—1968

In 1968, advances in technological research and studies, mainly geological studies, definitively launched Petrobras to offshore exploration. With these advances and new challenges, the need for new courses of improvement and changes in professionalization arises. However, surely, the process was a little different from the fifties, when CNP and Petrobras needed to prepare courses, select candidates, train and specialize skilled manpower for the oil industry.

Following the chronological classification established at the beginning of this book and communing the understanding of the chemical and processing engineer Jorge Naves Caldas, we can state that the years 1952–1965 "were characterized by the emphasis on the preparation of manpower for the emerging industry, because there were technicians in sufficient quantity and quality in Brazil" (Caldas 2005: 14). A second period, which began in 1966 and lasts until 1975, marks:

> [...] the beginning of the phase of absorption of more sophisticated technologies, which will generate the need, besides the basic training in the human resources of the company, of specialization and deepening in different areas. It is the time of searching for new exploration frontiers, towards activities at sea (Caldas 2005: 14).

From this, we elaborated the Organization Chart 4.6, which presents the courses and internships that it was intended to offer in 1968. With this, we closed the period of approach of the book, as, in that year, it occurs "the discovery of the first offshore well [...] in the field of Guaricema (Sergipe-Alagoas Basin), and the first drilling also in 1968, in the Campos Basin, in the field of Garoupa (Neto and Costa 2007: 100). In the same decade, Brazil would discover its largest oil reserves and again invest in building its deepwater know-how.

[29]Figures 4.14, 4.15, 4.16, 4.17, 4.18 and 4.19 Belong to the Exploração Na ... (S.D.).

Fig. 4.14 Drilling I

Fig. 4.15 Drilling II

Fig. 4.16 Drilling III

Fig. 4.17 Drilling IV

Fig. 4.18 Drilling V

Fig. 4.19 Drilling VI

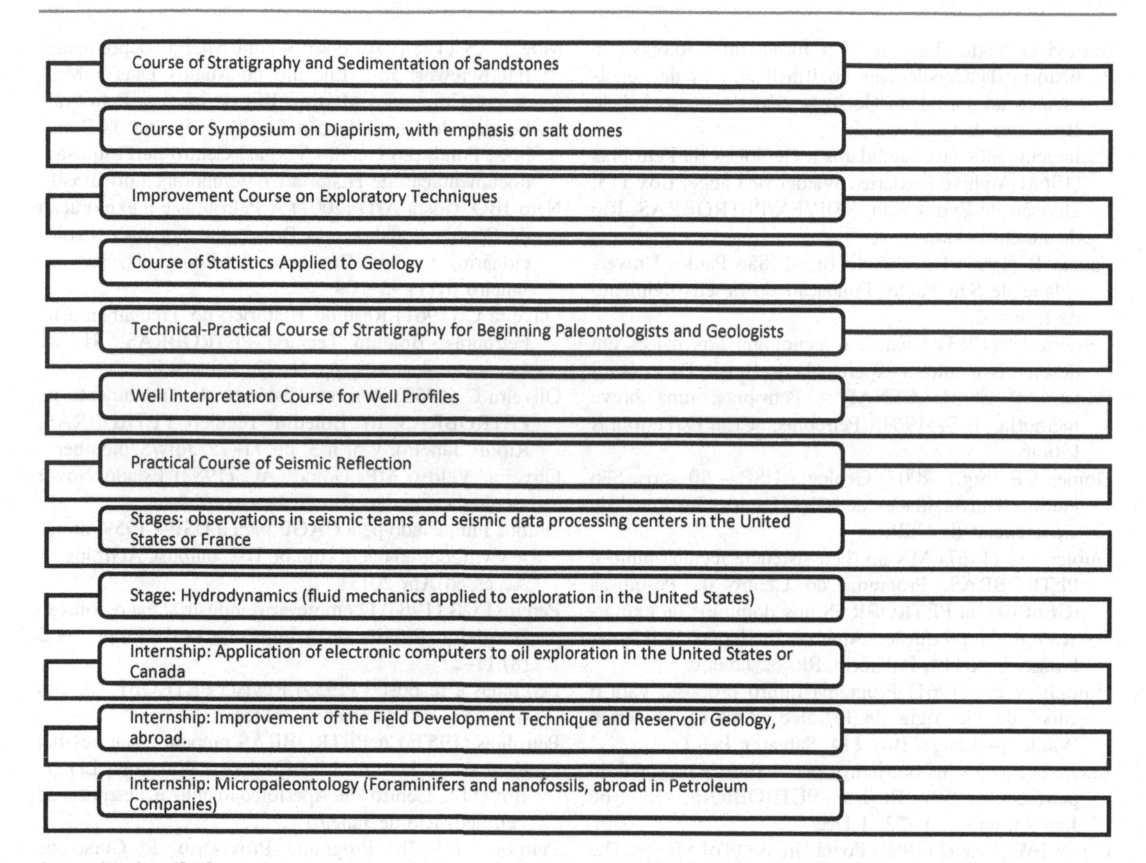

Organizational Chart 4.6 Courses and internships predicted for 1968 (Elaborated from Petrobras (1967b)). *Source* Elaborated by the author

References

Azevedo RLM, Terra GJS (2008) A busca do petróleo, o papel da Petrobras e o Ensino da geologia no Brasil. Boletim de Geociências da Petrobras. Rio de Janeiro 16(2):373–410

Beer JJ, Lewis WD (1963) Aspects of the Profissionalization of Science. The MIT Press on behalf of American Academy of Arts & Science, Daedalus 92 (4):764–784

Brasil. Decreto nº 40.783, de 18 de janeiro de 1957. http://legis.senado.leg.br/legislacao/listaPublicacoes.action?id=171995&tipodocumento=dEC&tipoTexto=PUB. Accessed Jan 2013

Caldas JN (2005) Uma história de sucesso: 50 anos de desenvolvimento de recursos humanos. Rio de Janeiro: PETROBRAS

Castells M (1999) A Sociedade em Rede. Paz e Terra, São Paulo

CENTRO DE APERFEIÇOAMENTO E PESQUISA DA PETROBRAS – CENAP. Normas reguladoras do Curso de Revisão. 1957a. CEnPEs/Petrobras Library

CENTRO DE APERFEIÇOAMENTO E PESQUISA DA PETROBRAS – CENAP. Curso de Geologia – Programa Provisório para o período de 1º de abril a 30 de junho. [s. l.: s.n.], 1957b. Archive Frederico Waldemar Lange, Box 110

Crisolis Reyes EC (2011) Los desafios técnicos y tecnológicos de la expropiación petrolera en México: El papel del Estado y la comunidad científica y tecnológica. Thesis. Facultad de Filosofía y Letras, UNAM, México

Dias J de NT (1991) José de Nazaré Teixeira Dias: Interview [1988]. Rio de Janeiro, CPDOC/FGV – SERCOM/ Petrobrás, p 374

Encontro de geólogos (1966) Porto alegre. Anais…. Porto alegre: Universidade Federal do rio Grande do Sul/Escola de Geologia. Sob patrocínio do Conselho Nacional de Pesquisas

Entrevista Viktor Leinz (1982) Jornal do Geólogo – a história da Geociências no Brasil através de depoimentos ao Jornal do Geólogo. São Paulo, sociedade Brasileira de Geologia
Esclarecimentos aos candidatos a Geólogos da Petrobrás (1965) Archive Frederico Waldemar Lange, Box 113, Divisão de Exploração – DIVEX/PETROBRAS, Rio de Janeiro, October
Fausto B (1995) História do Brasil. São Paulo: Universidade de São Paulo; Fundação do desenvolvimento da Educação
Ferreira JP (1983) Ciência e tecnologia nos países em desenvolvimento: a experiência do Brasil. Paris: [s.n.]
Fortes AP (2003) CENAP – Petrobras: uma breve memória 1954–1964. Petrobras, CEnPEs/Petrobras Library
Gomes CB (org.) (2007) Geologia USP – 50 anos. São Paulo: Universidade de São Paulo; Instituto de Geociências da USP
Gubler YG (1967) Missão de assistência técnica junto à PETROBRÁS. Programa do Centro de Pesquisas (CEnPEs) da PETROBRÁS nos domínios da exploração e da produção. Archive Frederico Waldemar Lange, Box 114, Relatório. Rio de Janeiro
Humphrey FL (1961) Programa futuro proposto para o curso de Geologia de Petróleo. Archive Frederico Waldemar Lange, Box 114, Salvador [s.n.]
Interesse dos jovens brasileiros pela indústria nacional do petróleo (1959) Revista PETROBRÁS. Rio de Janeiro, ano 6, n 153, 1 Dec
Lange FW [Letter] (1962) Ponta Grossa [To] Meijer. The possibility of employment with Petrobras. Archive Frederico Waldemar Lange, Box 79
Leal LRB, Leão IV (2008) Geologia na Bahia: 50 anos de história e desafios para a sociedade do futuro. Boletim de Geociências da Petrobras 16(2)
Ludwig G (1964) Nova Divisão Estratigráfica e Correlação Faciológica por meio de pequenas estruturas internas dos sedimentos silurianos e devonianos na Bacia do Médio Amazonas. Archive Frederico Waldemar Lange, Box 118, Centro de Aperfeiçoamento e Pesquisa de Petróleo – CEnaP; departamento de Exploração, Rio de Janeiro
Machado MM, Garcia IT (2013) Passado e presente na formação de trabalhadores jovens e adultos. Revista Brasileira de Educação de Jovens e Adultos 1(1):45–64
Mattoso S (2012) O curso de Geologia da Petrobras. Interview by Drielli Peyerl
Medina Pena L (2004) La invención del sistema político mexicano. Fondo de Cultura Econômica, México
Moggi AS [Carta] (1957) [To] Sr. Geólogo-Chefe do DEPEX. Lecture Prof. Lange. 1 f
Moggi AS [Letter] (1961) Rio de Janeiro [To] Aguiar, Manoel Pinto de aguiar. 2 f. Archive Frederico Waldemar Lange, Box 114
Moggi AS (1968) Pessoal para o avanço tecnológico – a experiência da PETROBRÁS. Jornal Diário de Notícias, Archive Frederico Waldemar Lange, Box 114, Rio de Janeiro, January
Moggi AS (1988) Antonio Seabra Moggi: depoimento. Interbviewer: José Luciano de Mattos Dias e Margareth Guimarães Martins. Rio de Janeiro, Petrobrás; Serviço de Comunicação Social; Memória da Petrobrás; Fundação Getúlio Vargas; Centro de Pesquisa e documentação de História Contemporânea do Brasil
Neto JBO, Costa AJD (2007) A Petrobrás e a exploração de Petróleo offshore no Brasil: um approach evolucionário. Revista Brasileira de Economia, Rio de Janeiro 61(1):95–109
Oliveira C (1961) Resumo Histórico do Treinamento na Petrobrás. Boletim Técnico PETROBRÁS, Rio de Janeiro, v 4, n 1–2, pp 71–72, January/June
Oliveira C (1962) Resumo Histórico do Treinamento na PETROBRÁS III Boletim Técnico PETROBRÁS, Rio de Janeiro, v 5, n 3, pp 71–72, July/September
Oliveira, Veloso MP, Gomes AC (1982) Estado Novo: ideologia e poder. Rio de Janeiro: Zahar
Ônibus Financiado pela CAGE para a USP—1959. http://www.figueiradaglete.com.br/T62_onibusCAGE.jpg. Accessed Apr 2013
Pereira LCB (1963) O empresário industrial e a revolução Brasileira. Revista de Administração de Empresas 2 (8):11–27
Perguntas e respostas (1959) Revista PETROBRÁS. ano 6, n 153, Rio de Janeiro, 1 Dec
Petrobras (1957a) A PETROBRÁS prepara o seu pessoal técnico: manual. Archive Frederico Waldemar Lange, Box 114, Centro de Aperfeiçoamento e Pesquisa de Petróleo, Rio de Janeiro
Petrobras (1957b) Programa Provisório do Curso de Introdução a Geologia. [s. l.: s.n.]. Archive Frederico Waldemar Lange, Box 110
Petrobras (1957c) Formação e Aperfeiçoamento de pessoal. Report. Archive Frederico Waldemar Lange, Box 114, Diretoria Executiva da Petróleo Brasileiro S. A PETROBRÁS,Rio de Janeiro
Petrobras (1957d) Resolução n° 7/57. Formação e aperfeiçoamento de Pessoal. In: Report. Archive Frederico Waldemar Lange, Box 114, Rio de Janeiro [s.n.]
Petrobras (1959a) Centro de aperfeiçoamento e Pesquisa de Petróleo. Curso de Manutenção de equipamentos de petróleo: manual. Archive Frederico Waldemar Lange, Box 114, Rio de Janeiro [s.n.]
Petrobras (1959b) Centro de aperfeiçoamento e Pesquisa de Petróleo. Curso de Refinação de Petróleo. Archive Frederico Waldemar Lange, Box 114, Rio de Janeiro [s.n.], pp 7–9
Petrobras (1960) Relatório preliminar do Grupo de Trabalho instituído pela Resolução n° 25/60, da Diretoria Executiva, para estudo da criação de um órgão de pesquisas para a indústria do petróleo. Archive Frederico Waldemar Lange, Box 114, Rio de Janeiro [s.n.]
Petrobras (1963) Centro de aperfeiçoamento e Pesquisa de Petróleo. Curso de engenharia de petróleo. CENPES/Petrobras Library, Rio de Janeiro [s.n.]
Petrobras (1966) Manual de Geologia de Superfície. N. 2. Archive Frederico Waldemar Lange, Box 14, Centro

de Pesquisas e Desenvolvimento (CENPES), Departamento Industrial (DEPIN), Rio de Janeiro

Petrobras (1967a) Centro de Pesquisa e desenvolvimento. dez anos de Evolução Tecnológica. Boletim Técnico da PETROBRÁS, Archive Frederico Waldemar Lange, Box 113, Rio de Janeiro, v 10, n 1, January/March

Petrobras (1967b) Relatórios Mensais. Archive Frederico Waldemar Lange, Box 32

Petrobras vai contratar técnicos francêses para ver se petróleo existe (1961) Jornal Diário de Notícias. 36, n 252, Porto Alegre, 1 Jan

Petróleo (1855) O Auxiliador da Industria Nacional. N 4, October

Pinto ID (2011) Interview - Irajá Damiani Pinto. By Drielli Peyerl. Porto alegre: [s.n.], Aug 2011

Primeiros geólogos diplomados no Brasil. Jornal do Estado de Minas, 1960. http://www.mmm.org.br/media/usuarios/511/imagens/geologos3.jpg. Accessed 14 May 2013

Recrutamento de técnicos para Landulpho Alves (1961) Revista PETROBRÁS, Rio de Janeiro, ano 7, n 184, June

Recrutamento e seleção de candidatos aos cursos do CENAP (1957) Boletim Técnico da PETROBRÁS, Rio de Janeiro, ano 1, n 1, October

Reunião de estudos técnicos de petróleo, between 5 and 11 October, 1958. Archive Frederico Waldemar Lange, Box 115, Bahia, Petrobras, 1958

Rezende F (2010) Planejamento no Brasil: auge, declínio e caminhos para a reconstrução. Brasília, DF: CEPAL/IPEA, 2010

Salles-Filho S (2000) Ciência, tecnologia e inovação – a reorganização da pesquisa pública no Brasil. Campinas: Editora Komedi

Sial AN (2008) Cinquenta anos de Geologia em Pernambuco (1957–2007): retrospectiva. Boletim de Geociências da Petrobras 16(2)

Silva DE (1952) A questão do petróleo. Rio de Janeiro: [s.n.]

Williams FC et al (1960) Boletim Técnico da Petrobrás. Archive Frederico Waldemar Lange, Box 48, CENAP, v 3, n 2, Rio de Janeiro, pp 161–166, April/June

5 Conclusions

Among the various works in this book, we find diversified points of approach and divisions about the history of oil in Brazil: visions directed to the nationalism (Cohn 1968), politics, and economics (Moura and Carneiro 1976; Smith 1978; Júnior 1989), among others.

In fact, the purpose of this book was to describe and demonstrate the various paths taken in the search and exploitation of oil in Brazil, which led to a destiny: the necessary, indispensable, and urgent formation of professionals to meet the need of manpower in the sector. It is at this point that the National Petroleum Council and Petrobras became institutions[1] providing their own manpower, while the teaching conditions in the country were deficient in the area of geosciences, fomenting the formation of a network of foreign and Brazilian professionals.

To achieve the conclusions set out below, the structure of the book was based on studies of the History of Science, demonstrating science and scientific practice in the peripheral countries[2] (Lafuente and López-Ocón 1998) and how rich was its development, specifically in Brazil, for geological sciences.

We began our journey in 1864, with an official decree that brings the word *petroleum* for the first time in Brazilian law. From that on, we seek to highlight the process of development of technical–scientific research for the exploration of oil in the national territory, a process that was surrounded by private and governmental initiatives, demonstrating the existence of scientific

[1]"As an institution, we understand a specific place that serves as a transit channel and discussion among people, knowledge, artifacts and ideologies and has as its ultimate purpose the elaboration, verification and continuation of scientific knowledge as legitimate and true. A historical analysis of the institutionalization process necessarily involves the identification of these various aspects, such as mapping, and the monitoring of these elements over time, which proposes a historical interpretation. The interpretation comprises the relation of this set of data between themselves and between them and other local or general historical contexts, which are also of a different nature from the scientific one. In this way, it is expected to understand the ways or mechanisms by which that set of knowledge was institutionalized, both in the scientific community and in the society in general" (Oliver and Figueirôa 2006: 105).

[2]"It is possible to identify peripheral science not only in countries or regions outside the world centers of power, but also in central countries, if the criteria are not satisfied. I repeat that although the term peripheral science is sometimes used pejoratively, it has no such connotation here, and should be understood simply within what has been established. Many cases of peripheral science of very high quality and relevance can be pointed out. Sometimes, from a scientifically peripheral country, notable contributions can come to central science; this can be seen as evidence that, despite being on the periphery, that society has great scientific potential, which may or may not manifest itself, depending on several factors. This aspect, by itself, would already justify the need to know the history of peripheral science" (Filgueiras 2001: 710).

D. Peyerl, *The Oil of Brazil*, Historical Geography and Geosciences,
https://doi.org/10.1007/978-3-030-13884-4_5

activities that contributed to the development of the technique and for the training of manpower related to the geological sciences, precisely the geology of petroleum.

When performing the first deep explorations, we observed the presence of foreigners and the use of technology from abroad (such as the importation of drills, for example). This first phase, which we demonstrated in the first part of the book, describes the absorption of knowledge by the empirical method and by means of manuals, elaborated in English and French, which at first made it difficult to improve and to apply the technique for exploration.

Another problem faced by Brazilians in the search for oil was the scarcity of geological knowledge in the period, which was tied to the extension of the territory and the absence of advanced technology. The discovery of oil in other countries, through wild cat explorations, corroborates that the knowledge about where to look for oil was in full transition. The case of Mexico, which, in the first decades of the twentieth century, found greater occurrences of oil in geological dispositions where, by the American criteria, there would be no evidence of it (Pedreira 1927).

Negative data from reports of foreign professionals, such as the White report (1908), did not discourage Brazilian professionals—trained in Natural Sciences or, mainly, by the School of Mines of Ouro Preto in Mining and Civil Engineering, self-taught and specialized abroad—in seeking oil in the country.

Brazil began the oil exploration with professionals from different areas, mainly from the geological and natural sciences, who dedicated themselves to the studies of the area and produced precise information about the geology of the national territory.

Due to factors and events such as the First World War (1914–1918), oil became one of the world major sources of energy, and the Brazilians, because of the territorial conditions and the extension of our country, believed it was possible to find black gold in abundance—after all, Latin American countries such as Peru and Venezuela had found oil in great quantity in this period. With this, the exploration of oil in Brazil could still be possible, it simply needed to be found.

After the war period, both the federal government and some state governments began to deliberately stimulate the development of some specific and diversified industries (Suzigan 2000). This eventually leverages the process of industrialization and arouses a huge interest in oil, in parallel to the gradual replacement of the coal.

Thus, after overcoming the difficulties found, and starting from the solutions adopted, the first oil discovery was made in Brazil, in the district of Lobato of Bahia, in 1939.

The process until this happened was marked by insistence and perseverance. The main motivations were political and economic petroleum issues related to the presence of foreign companies in the country, the presence of private initiatives, and, finally, the consolidation of governmental initiatives. The modifications made over the years culminated in the creation of the National Petroleum Council (1938).

The apparent mismatch of failed attempts to search for oil in the first part of this book has resulted in initiatives that have led to the discovery of the first exploratory well, a proper oil policy, and investment in geological sciences related studies. Perhaps such mismatch is due to the recurrent investment in the same places of previous exploratory attempts in search of oil. We point out that such conduct was justified by the technique previously applied and by geological knowledge of the territory, which was insufficient for the period during which the explorations took place. The improvement took place only over the years, including the creation of a technical–scientific network made up of Brazilians and foreigners.

The aforementioned commissions and initiatives have given place to the creation/integration of a collection of oil exploration data in the country that served as a starting point for the organization of exploration activities of the National Petroleum Council and Petrobras.

In the second part, we presented the work developed by the National Petroleum Council and by Petrobras—through the Exploration

Department (DEPEX)—in the intersection between 1938 and 1961, focusing on established relations between Brazilians and foreigners to build their own know-how.

The creation of the CNP establishes its own petroleum policy, nationalizing its industry even before it is discovered. In this period, nationalist issues, already present in previous years, including laws, take precedence over oil. Despite that, the absence of Brazilian skilled labor in the sector required the permanent presence of foreigners for the development and improvement of the technique in the oil industry.

In addition, there is a shortage of resources directed to the CNP for the exploration and the formation of manpower, causing it, initially, to conduct its activities with the help of contracted foreign companies that could train Brazilians with technical and practical training. The budget also allowed sending Brazilians to internships abroad. However, these two attempts only softened the problems that CNP faced in relation to its goal, which was to find more oil producing wells.

Among the difficulties faced by the CNP in the exploration of new wells were the requirement of too much effort from the professionals and the precarious conditions to reach some places to start the research: many professionals—true pioneers of the national territory—were affected by local diseases and dehydration, and the transport of materials to carry out the activities was a factor of choice for places where studies should be carried out for the exploitation of oil, often not being economically feasible the expenditure of expenditures for the quantity of barrels extracted from a well. In addition, the CNP also suffered some political pressure from the presence of foreign companies in the country, which for many were here only to steal national oil—although in fact it had not yet been found in commercial quantity.

The search for oil is time consuming and it can be considered an arduous geological adventure, in which there is no guarantee of success (Peixoto and Peixoto 1957). In general, "the petroleum research includes: geological reconnaissance of the areas to be surveyed; geophysical works for the selection of drilling points; carrying out drillings to verify the existence of petroleum" (Peixoto and Peixoto 1957: 275). The constitutions and laws approved of this period until 1946 sought to extirpate any participation of foreigners in the country, in anything related to petroleum. Despite this, they remained very present because of the need for exploratory techniques, geological knowledge and their teaching for the formation of Brazilians.

Petrobras monopolized oil in 1953, and its investments were mainly directed toward exploration. Even so, it was necessary to structure its Exploration Department by appointing foreigners to the top positions, as Brazil still did not have self-sufficiency in the technique. The intention was, however, to replace them by Brazilians, when possible.

At that moment, the unique figure of Walter Link stands out, who gathered geological information from the Brazilian territory, being able to determine more precisely the oil probabilities in the country and suggesting the direction of the research for the continental shelf.

By hiring technicians (mainly geologists) from different countries, Petrobras adopted the training tactics of Brazilians through foreign professionals, starting to structure courses of improvement and professionalization.

The year 1961, when the second part of this book ended, is marked by the greater insertion of Brazilian professionals in the market, including oil market. However, Petrobras continued investing in the improvement and professionalism of the workforce in accordance with the demands that arose; this was reported in the last part of the book.

This reveals what we propose since the beginning: how CNP and Petrobras became providers of the Brazilian manpower, a need diagnosed since the end of the nineteenth century and which, after numerous attempts, was consolidated through two governmental institutions.

Petrobras, which instituted most of the training courses, became the object of study because it is one of the few known cases in which a mixed economy company assumes the responsibility of forming its own team, also contributing to the opening of an undergraduate course (Geology) and several postgraduate courses related to petroleum. It is evident, therefore, that Petrobras accelerated and contributed decisively to the creation of the Geology course in the country.

The reasons that led the company to continue investing in courses over the years are the following two: (1) it would not be possible for a geology course to focus only on oil-related issues (noting that both EMOP and later universities such as the Federal District and the Polytechnic School of the University of São Paulo turned their courses primarily to mineral exploration); and (2) Petrobras need for manpower increased according to the discovery of oil and with the development of new techniques, such as drilling at sea.

The work carried out by CNP and Petrobras led to exploratory research in Brazil and to investment in improvement and professionalization, obtaining more concise exploration data in search of self-sufficiency.

The external contributions were essential for the satisfactory development of the teaching offered by Petrobras. This work, carried out jointly with other scientific societies (such as the Brazilian Society of Paleontology) and with universities, contributed to the process of institutionalization and professionalism presented.

The disciplines of the courses offered by CNP and Petrobras contextualized different economic moments and national politicians, demonstrating elements that contributed to the improvement of future directives in the technical training of the company, such as the creation of the Petroleum Improvement and Research Center (CENAP) in 1955, characterized by the pioneering spirit already in its conception, not only for promoting numerous courses but also for implementing technological research. The activities of the Research and Development Center Leopoldo Américo Miguez de Mello (CENPES) in 1966 were also highlighted. Its purpose was to intensify and deepen the know-how available in the technical and scientific field of petroleum.

What we can observe in the course of this book is the formation of a technical–scientific network between Brazilians and foreigners focused on the development of petroleum exploration techniques in the country, and especially the courses offered by CNP and Petrobras. This network, formed by a range of scientists, self-taught researchers, politicians, military, engineers, Brazilian and foreign professionals, "nationalists" and "submissive", with different approaches and proposals, operated in the search of a common interest: Petroleum.

Among the various works quoted in this book, one tried to demonstrate, through the use of some works and sources of the period (it was not possible to cite all the works known and/or read, not all the names that were part of that process), the efforts that were being in the pursuit of oil and how much this stimulated permanent initiatives in the resumption and development of research. Such initiatives have encountered a great challenge, perceived since the beginning of our research, related to the several attempts to adapt the technology to the local geological conditions. For this barrier to be broken, it was necessary for the joint work of foreigners and Brazilians. Countries such as Mexico went further, having expropriated all foreign companies in 1938, nationalizing the oil industry, forming and maintaining, until the present day, the "powerful Petróleos Mexicanos—PEMEX" (Moura and Carneiro 1976). However, as demonstrated in this book, their professional training bases were different, although at first, they went through a similar case to that of Brazil as regards the need to build their own know-how in the face of staff shortages qualified.

Nationalism in Brazil apparently banned the entry of foreign companies. On the other hand, it contributed to the state monopoly. Under these two aspects, we affirmed not only that there was no consolidation of foreign companies in the country but we add that the development of the exploratory technique and professional training in the country was the result of the joint work of Brazilians and foreigners.

However, there is still a pause for the technical development in the area of Paleontology by Petrobras, which contributed significantly to the development and application of the technique for the discovery of new oil wells and/or exploration sites and for the area. From 1957 onwards, the first systematic work on the analysis of pollen (Palynology) concentrated in continental formations of the Recôncavo Basin. In 1960, the studies of ostracods introduced in Bahia State and later in Belém city are mentioned, the studies of foraminifera started in Maceió City, later in Belém City and Ponta Grossa City, and, in parallel, the study of organized groups, chitinozoal (Micropaleontology[3]) (Gubler 1967), and others. New methods were also introduced, such as the examination of nannofossils (for the study of Cretaceous and Tertiary marine formations). We closed the emphasis given to the area of paleontology with the ideas of the Diogenes is paleontologist of Almeida Campos, who stated that the important need to find oil in Brazil provoked and stimulated paleontology in the country (Diógenes de …, s.d.).

Brazil succeeded in its project through CNP and, mainly, Petrobras, which invested in the construction of its own know-how, evidencing itself as one of the only companies to create courses for the formation while the industry was in wide expansion. Using the foreign technology and forming a workforce capable of operating it without dependence on any of its original planners and designers, the time in the assembly of an industrial plant related to oil became shorter (Oliveira 1961). For this reason, CNP and Petrobras have become one of the objects of study in this book.

We ended the book in 1968, when Petrobras directed part of its research to the sea with the discovery of the first offshore well in the field of Guaricema (Sergipe State), initiating a new cycle of technological knowledge of exploration. Most of the imported technology for exploration of the period was American, given that the large oil reserves in other countries—such as the United States of America—were located in territorial sedimentary basins, that is, most of the knowledge and technological development did not have as main investment the offshore research. Moreover, "the little of the technological knowledge of offshore oil exploration of the time, also did not fit the Brazilian reality" (Neto and Costa 2007, p. 96).

> Faced with such a technological impasse, the Brazilian authorities had to decide between producing a technology in keeping with the local reality; acquire such technology via a contract with international institutions; or else import the mineral. Perhaps influenced by nationalist military consciousness, the strategic importance of the country natural resources, as well as the lack of international know-how, the decision was to produce locally a system of innovations that would allow the exploration of offshore oil, a technology known as offshore. Whatever the motivation of this decision, Petrobras, through its Deepwater Technology Training Program – PROCAP – created in 1986, has been following a path of several discoveries, which has given the institution the title of international leader in exploration technology of deep-water oil (Neto and Costa 2007: 100).

Thus, Petrobras was able to establish a minimum of know-how to consolidate the company as technically autonomous, with the exception of geophysics, which continued to depend on foreign companies. In relation to the training of professionals and the proposal and the performance of a training program, the pioneering form as these were put into practice by CNP and Petrobras is worthy of encumbers, contributing to the formation and consolidation of areas geared towards geological sciences in Brazil.

[3]Frederico Waldemar Lange was the pioneer in the development of the technique for the study of Micropaleontology in Brazil (Peyerl 2010).

References

Cohn G (1968) Petróleo e nacionalismo. Difusão Européia do livro, São Paulo

Filgueiras CAL (2001) A história da ciência e o objeto de seu estudo: confrontos entre a ciência periférica, a ciência central e a ciência marginal. Revista Química Nova, são Paulo 24(5)

Gubler YG (1967) Missão de assistência técnica junto à PETROBRÁS. Programa do Centro de Pesquisas (CEnPEs) da PETROBRÁS nos domínios da exploração e da produção. Archive Frederico Waldemar Lange, Box 114, Relatório. Rio de Janeiro

Júnior IPM (1989) Petróleo: política e poder: um novo choque do petróleo? José Olympio, Rio de Janeiro

Lafuente A, Lopéz-Ocón L (1998) Bosquejos de la ciencia nacional em la américa latina del siglo XiX. Asclepio – Revista de historia de la medicina y de la ciência 2(2):5–10
Moura P, Carneiro FO (1976) Em busca do petróleo brasileiro. Ouro Preto: Fundação Gorceix
Neto JBO, Costa AJD (2007) A Petrobrás e a exploração de Petróleo offshore no Brasil: um approach evolucionário. Revista Brasileira de Economia, Rio de Janeiro 61(1):95–109
Oliveira C (1961) Resumo Histórico do Treinamento na Petrobrás. Boletim Técnico PETROBRÁS, Rio de Janeiro, v 4, n 1–2, pp 71–72, January/June
Oliver GS, Figueirôa SFM (2006) Características da institucionalização das ciências agrícolas no Brasil. Revista Brasileira de História da Ciência, Rio de Janeiro 4(2):104–115
Pedreira A de B (1927) A pesquiza de petroleo. Typographia do «Annuario do Brasil», Rio de Janeiro
Peixoto JB, Peixoto W (1957) Produção, transporte e energia no Brasil. [s.l.: s.n.]
Peyerl D (2010) A trajetória do paleontólogo Frederico Waldemar Lange (1911–1988) e a História das ciências. Thesis. State Ponta Grossa University. Ponta Grossa, 116 f
Smith PS (1978) Petróleo e política no Brasil Moderno. Artenova, Rio de Janeiro
Suzigan W (2000) Indústria brasileira: origem e desenvolvimento. Hucitec and Unicamp, São Paulo

Appendix A
Decrees and Federal Decree-Laws

1864

Decree No. 3.352-A, from November 30, 1864

Grants Thomaz Denuy Sargent a faculty for a period of 90 years, either by himself or through a company, to extract peat, oil, and other mines in the Comarcas do Camamu and Ilheos, of the province of Bahia.

1872

Decree No. 5.014, from July 17, 1872

Grants Luiz Matheus Maylaski permission for 2 years to explore stone coal and petroleum in the regions of Sorocaba, Itapetininga, and Itu, in the province of S. Paulo.

Decree No. 5.050, from August 14, 1872

Grants Dr. Cyrino Antonio de Lemos and Dr. João Baptista da Silva Gomes Barata, permission of 2 years to mine stone coal and oil in the region of the capital of the province of S. Paulo.

1874

Decree No. 5.732, from August 27, 1874

Prolongs for 1 year the period set for the Graduated Cirino Antonio de Lemos and José Baptista da Silva Gomes Barata, in the Decree No. 5.050 from August 14, 1872, for the exploration of coal and oil mines in the Comarca of the capital of province of S. Paulo.

1882

Decree No. 8.416, from February 11, 1882

Grants Antonio Lopes Cardozo a privilege for the process of his invention designed to make oil unexplained, disinfected, and colored.

Decree No. 8.703, from October 7, 1882

Grants permission to Gustavo Emilio Olander to explore oil fields in the regions of Campo Largo and Lapa, in the province of Parana.

1883

Decree No. 8.840, from January 5, 1883

Grants permission to Dr. Gustavo Luiz Guilherme Dodt and Graduated Tiberio César de Lemos to explore minerals [including oil] in the Maranhao Province.

Decree No. 8.983, from August 4, 1883

Extends the deadline granted to Antonio Lopes Cardoso for the process of his invention, designed to render kerosene or oil inexplosive.

1885

Decree No. 9.444, from July 20, 1885

Grants permission to Manoel Vidal Barbosa Lage to explore stone coal and oil in the province of Minas Gerais.

Decree No. 9.493, from September 5, 1885

Renews the concession referred to in decree No. 5744[1] from September 16, 1874 for exploration of stone coal and oil in the province of S. Paulo.

1887

Decree No. 9.724, from February 19, 1887

Grants Henri Raffard permission to transfer to Major Francisco de Assis Paula the concession referred in Decree no. 9493 from September 5, 1885.

[1]The Decree nº 5744, from 16 September 1874 mentions only the mining of stone coal from Água Branca. When the concession was renewed by the decree No. 9493, from 5 September 1885, oil exploration is added.

D. Peyerl, *The Oil of Brazil*, Historical Geography and Geosciences,
https://doi.org/10.1007/978-3-030-13884-4

1888

Decree No. 10.037, from September 15, 1888

Grants permission to Ignacio de Souza Lages to explore stone coal, oil, and other minerals in the city of Cameta, Para Province.

Decree No. 10.073, from November 8, 1888

Grants permission to Tito Livio Martins to explore oil and other minerals in the city of Tatuhy, province of S. Paulo.

Decree No. 10.105, from December 1, 1888

Extends for 1 year the period specified in the decree No. 9.724 from February 19, 1887, for the conclusion of the exploration work of oil and stone coal in the city of Tatuhy, province of S. Paulo.

1889

Decree No. 10.239, from May 2, 1889

Grants permission to João Maria do Valle, Engineer Abdon Felinto Milanez and Emilio de Menezes to explore coal stone, oil, and other minerals in the valleys of the *ribeirões of* Cannavieiras and Cubatao, in the municipality of Guaratuba, Parana Province.

Decree No. 10.347, from September 6, 1889

Grants permission to João Moreira da Silva to explore stone coal and other minerals, oil, and other bituminous substances in the province of Santa Catharina.

Decree No. 10.361, from September 14, 1889
Grants permission to Raulino Julio Adolpho Horn to explore oil and other mineral oils in the province of Santa Catharina.

Decree No. 10.431, from November 9, 1889

Grants permission to Adam Benaion to explore oil, stone coal, and other minerals in the province of Para.

Decree No. 10.445, from November 9, 1889

Grants to Tito Livio Martins an extension of the term established in Decree No. 10,073 from November 8, 1888.

1890

Decree No. 393, from May 12, 1890

Grants permission to Dr. Almir Parga Nina to explore, by himself or through a company, mineral oils (including petroleum) in the State of Maranhao.

Decree No. 670, from August 18, 1890

Grants permission to Tito Livio Martins to explore oil and other minerals in the city of Tatuhy, province of S. Paulo.

Decree No. 1.114, from November 29, 1890

Extends the deadline to Raulino Julio Adolfo Horn to explore oil and other mining oils in the State of Santa Catarina.

1897

Decree No. 2.471, from March 8, 1897

Approves, with amendment, the articles of the Petroleum Industrial Company and authorizes it to operate.

1912

Decree No. 9.335, from January 17, 1912

Grants the Standard Oil Company of Brazil, current denomination of the Petroleum Industrial Company, authorization to continue to operate in the republic.

1913

Decree No. 10.168, from April 9, 1913

Grants authorization to The Anglo Mexican Petroleum Products Company, Limited, to operate in the republic.

1917

Decree No. 12.438, from April 11, 1917

Grants authorization to "The Anglo Mexican Petroleum Products Company, Limited", to replace this denomination by "Anglo Mexican Petroleum Company, Limited".

1932

Decree No. 21.414, from May 17, 1932

Authorizes Brasileira de Petróleo Company to continue with the assignment and lease agreements for sub-of territorial properties in the municipality of Pirajú in the State of São Paulo.

Decree No. 21.415, from May 17, 1932

Authorizes the incorporation by Mr. J. B. Monteiro Lobato, M. L. de Oliveira Filho e L. A. Pereira de Queiroz of a joint-stock company with headquarters in São Paulo and capital of 3.000:060$0 (three thousand "contos de réis"), exclusively national, with the objective of researching oil formations and exploring the respective deposits.

Decree No. 22.210, from December 13, 1932

Authorizes Th. Marinho de Andrade, Augusto Leal de Barros, and Constantino Badesco Dutza to organize a company for the exploration of oil, with the denomination of National Company for Exploration of Oil.

1933

Decree No. 22.932, from July 12, 1933

Authorizes Avelino Barreto to hire, without privilege, research, and exploration of petroleum and asphalitis in the city of Botucatu, State of Sao Paulo, and to organize a company for the exploitation of the contract.

Decree No. 23.225, from October 17, 1933

Authorizes, without privilege, Companhia Brasileira de Petroleo Cruzeiro do Sul, to hire with Rita Spinola Dias, owner of Fazenda Bofete, in the municipality of Porangaba, and with Adelaide Barnsley Guedes, or its successors, owner of Fazenda Pederneiras, in the municipality of Tatui, both municipalities of the State of Sao Paulo, to the research and exploration of petroleum that exists in those farms.

Decree No. 23.572, from December 12, 1933

Authorizes Th. Marinho de Andrade, Augusto Leal de Barros, and Constantino Badesco Dutza to include the contracts they made with Rodolfo Jacob and Elói José Nuncio, in the areas dealt with in art. 1 of decree n. 22.210, from December 13, 1932, for the research and exploration of petroleum by admitting the company "Companhía Nacional para Exploração de Petróleo", in organization, and makes other provisions.

Decree No. 23.575 of December 12, 1933

Authorizes, without privilege, Companhia Brasileira de Petróleo, an anonymous society, with headquarters in the Federal Capital, to contract the acquisition or lease of territorial properties in the municipality of Ribeirao Claro, in the State of Parana, for the research and exploration of oil.

1934

Decree No. 23.752, from January 16, 1934

Authorizes its privilege, the Pan-Brazilian Petroleum General Company, with headquarters in the Federal Capital, to contract the acquisition or lease of territorial properties in the municipality of Ribeirao Claro, State of Parana, for the research and exploration of oil.

Decree No. 24.377, from June 12, 1934

Authorizes, without privilege, Companhia Brasileira de Petróleo, an anonymous society, headquartered in the Federal Capital, to:

(1) To contract the lease of lands belonging to private individuals, located in the municipality of reserve, in the State of Parana, in order to search for oil;
(2) Enter into the purchase option contracts for the aforementioned land; and
(3) Acquire the oil deposits that may exist in the subsoil.

1936

Decree No. 1.041, from August 20, 1936

Approves the relation of the personnel contracted for geological studies and petroleum researches in the Territory of Acre and State of Amazon.

1937

Decree No. 1.849, from August 3, 1937

Authorizes the Brazilian citizen Silvio Fróis Abreu to research natural gas and oil in an area of 175,84 hectares on the Ilha itaparica, in the municipality of Itaparica, State of Bahia.

Decree No. 1.850, from August 3, 1937

Authorizes Brazilian citizen Edgard Frias to search for oil and natural gas in private lands located in the district of Mapele, municipality of Matoim, State of Bahia.

Decree No. 1.870, from August 10, 1937

Authorizes the Brazilian citizen Silvio Fróis Abreu to research natural gas and oil in an area of 224,16 hectares on the Ilha Itaparica, in the municipality of Itaparica, State of Bahia.

Decree No. 2.119, from November 9, 1937

Authorizes the Brazilian citizens Olímpio José Brochado, Firmino de Santana, and Quineto Gusmão Rocha to search for oil on marine lands located in the place called "Porto de Sauípe", municipality of Entre Rios, State of Bahia.

Decree No. 2.189 of December 21, 1937

Authorizes, on a provisional basis, the Brazilian citizen Carlos Dias de Avila Pires, to search for oil and natural gas in the municipality of Monte Negro, State of Bahia.

Decree No. 2.190 of December 21, 1937

Authorizes, on a provisional basis, the Brazilian Society of Mineralogical Research Limited, to search for oil and natural gas, on the coast of the State of Bahia.

Decree No. 2.191 of December 21, 1937

Authorizes, on a provisional basis, the Brazilian Society of Mineralogical Research Limited, society organized in Brazil, to research oil and natural gas, on the coast of the State of Bahia.

Decree No. 2.192, from December 21, 1937

Authorizes, on a provisional basis, the Emprêsa Nacional de Investigação Geológicas Limitada, organized in Brazil, to research petroleum and natural gases on the coast of the State of Bahia.

Decree No. 2.193, from December 21, 1937

Authorizes, on a provisional basis, the Emprêsa Nacional de Investigações Geológicas Limitada, a company organized in Brazil, to research natural gas and oil in the Ilha Itaparica, municipality of Itaparica, State of Bahia.

Decree No. 89, from 21st December 1937

Approves the Special Protocol on Railway Connections and Use of Bolivian Petroleum signed in La Paz on November 25, 1937.

Decree No. 2.217, from December 28, 1937

Authorizes, on a provisional basis, the Brazilian citizen Salvador Prioli Junior by himself or a company to organize, to research oil and natural gas, in the State of Sergipe.

1938

Decree No. 366, from April 11, 1938

Incorporates to the Code of Mines, decree No. 24.642, of July 10, 1934, new title, which establishes the legal regime of deposits of oil and natural gases, including rare gases.

Decree No. 380, from April 18, 1938

Approves the Treaty on the Exit and Use of Bolivian Petroleum, between Brazil and Bolivia, signed in Rio de Janeiro, on February 25, 1938.

Decree No. 395, from April 29, 1938

Declares as public utility and regulates the importation, exportation, transportation, distribution, and trade of crude oil and its derivatives in the national territory, as well as the refinery industry of imported oil produced in the country, and other measures.

Decree No. 2.616, from May 4, 1937

Authorizes, on a provisional basis, the Brazilian citizen Alberto Hofmann, by himself or a company to organize, to search for oil in Serra da Taquara Verde region, in Rio Caçador municipality, in the state of Santa Catarina.

Decree No. 2.800, from June 29, 1938

Authorizes, the provisional title, the Brazilian citizen Francisco de Sá Lessa for himself or society to organize, to search for oil and natural gas in the municipality of Alagoas, State of Alagoas.

Decree No. 2.801, from June 29, 1938

Authorizes the Brazilian citizen Tadeu de Araújo Medeiros for himself or society to organize research of oil and natural gas in the municipality of Alagoas, State of Alagoas.

Decree No. 2.802, from June 29, 1938

Authorizes, on a provisional basis, the Brazilian citizen Eudoro Lemos de Oliveira, by himself or a company to organize, to search for oil and natural gas in the municipality of Coruripe, State of Alagoas.

Decree No. 2.803, from June 29, 1938

Authorizes, on a provisional basis, the Brazilian citizen of Barbosa de Couto e Silva, by himself or a company to organize, to search for oil and natural gas in the municipalities of Socorro and Laranjeiras, State of Sergipe.

Decree No. 2.804, from June 29, 1938
Authorizes, on a provisional basis, the Brazilian citizen Eurico de Rocha Portela, by himself or a company to organize, to search for oil and natural gas in the municipality of Maceio, State of Alagoas.

Decree No. 2.805, from June 29, 1938
Authorizes, on a provisional basis, the Brazilian citizen Cristiano Heyn Hamann, for himself or a society to organize, to search for oil and natural gas in the municipality of Piassub-ussu, in the state of Alagoas.

Decree No. 2.806, from June 29, 1938
Authorizes, on a provisional basis, the Brazilian citizen Oscar Edvaldo Portocarreiro, by himself or a company to organize, to research oil and natural gas in the municipality of Santo Amaro, State of Sergipe.

Decree No. 538, from July 2, 1938[2]
Organizes the National Oil Council, defines its assignments and other measures.

Decree No. 533, from July 5, 1938
Extends the term referred to in § 1 of art. 4th of decree-law No. 395, of April 29, 1938.

Decree No. 538, from July 7, 1938
Organizes the National Oil Council, and defines its assignments and other measures.

Decree No. 2.999, from August 17, 1938
Authorizes the Brazilian citizen, Elpidio Domingues Lins, to search for oil and natural gas in the municipality of Recife, capital of the state of Pernambuco.

[2]There is a record in the database of the Federal Senate website that the same decree-law no. 538 (which creates the National Oil Council) had two editions, one of 2nd July 1938, published in the Collection of the Brazil on 31st December 1938 and the other, dated 7th July 1938, published in the Federal Official Gazette on 8th July, 1938. Since the access to the contents of the first document is not available, it is not possible to know if it is identical to the one edited later (accessed). It was decided to keep records of the existence of the two documents in order to demonstrate the power of the State when the National Oil Council was created.

Decree No. 3.004, from August 19, 1938
Declares without effect the decree No. 2,616, from May 4, 1938, provisionally authorizing the Brazilian citizen Alberto Hofmann, by himself or a company to organize, to search for petroleum in the region of "Serra da Taquara Verde", in the municipality of Rio Caçador, in the state of Santa Catarina.

Decree No. 3.008, from August 19, 1938
Authorizes, on a provisional basis, the Brazilian citizen Alberto Hofmann, by himself or a company to organize, to search for oil in Serra da Taquara Verde region, in Rio Caçador municipality, in the state of Santa Catarina.

Decree No. 3.097, from September 22, 1938
Authorizes, on a provisional basis, the Brazilian citizen Vitor Amaral Freire, by himself or by the "Companhia Matogrossense de Petróleo", in organization, to search for oil and natural gas in borderlands located in the municipality of Corumba, in the State of Mato Grosso.

Decree No. 3.098, from September 22, 1938
Authorizes, on a provisional basis, the Brazilian citizen Vitor Amaral Freire, by himself or by the "Companhia Matogrossense de Petróleo", in organization, to search for oil and natural gas in borderlands located in the municipality of Corumba, in the State of Mato Grosso.

Decree No. 3.099, from September 22, 1938
Authorizes, on a provisional basis, the Brazilian citizen Vitor Amaral Freire, by himself or by the "Companhia Matogrossense de Petróleo", in organization, to search for oil and natural gas in borderlands located in Porto Esperança in the municipality of Corumba, in the State of Mato Grosso.

Decree No. 747, from September 29, 1938
Opens, by the Ministry of Finance, the special credit of 750:000$000 for the National Petroleum Council.

Decree No. 3.131, from October 05, 1938
Promulgates the Treaty on the Exit and Use of Bolivian Petroleum, between Brazil and Bolivia, signed in Rio de Janeiro, on February 25, 1938.

Decree-law No. 804, from October 24, 1938

Extends the term referred to in the sole paragraph of art. 3 of decree-law n. 395, from April 29, 1938.

Decree-law No. 842, from November 9, 1938

It fixes the salaries of the members of the Executive Committee of the National Petroleum Council and takes other measures.

Decree No. 3.344, from November 30, 1938

Declares as expired the authorization granted to Avelino Barreto, by decree no. 22,932, dated July 12, 1933, to contract, with no privileges, the exploration and exploitation of oil and asphalitis in lands located in the city of Botucatu, State of Sao Paulo, as well as to organize a society for the exploration of the contracts to be carried out.

Decree No. 3.452, from December 14, 1938

Declares as expired the authorization granted to Olimpio José Brochado, Firmino de Sant'anna and Quineto Gusmão Rocha, by decree number 2, 119, from November 9, 1937, to search for oil on land located in the place called Porto de Sauípe, in the Municipality of Entre Rios, Bahia State.

Decree-law No. 961, from December 17, 1938

Rewrites No. I of art. 3 of decree-law n. 395, from April 29, 1938.

Appendix B
Table B.1

Table B.1 Oil drillings made by the Federal government all over national territory from 1919 to 1928

SONDAGENS PARA PETROLEO EFETUADAS PELO GOVÈRNO FEDERAL EM TODO O TERRITORIO NACIONAL, de 1919 a 1928					
Estados	Localidades	Encarregados	Início	Fim	Prof. metros
(1) Paraná	Marechal Mallet	R. Lima Coelho	18-8-19	7-7-20	84, 77
(2) Alagoas	Garça Torta	A. Bulhões Pedreira	14-4-20	9-7-20	130, 26
(3) Paraná	Marechal Mallet	Alpheu Diniz Gonsalves	21-9-20	23-3-22	509, 97
(4) Alagoas	Garça Torta	A. Bulhões Pedreira	26-11-20	24-3 21	130, 26
(5) Bahia	Ilhéus, Cururipe	Gerson Alvim	24-2-21	8-4-21	100, 51
(6) Bahia	Ilhéus, Cururipe	Júlio Põrto	1-6-21	14-11-21	197, 05
(7) São Paulo	Graminha, São Pedro	Gerson Alvim	12-2-21	21-7-22	329, 43
(8) São Paulo	Querozene, São Pedro	Bourdot Dutra	5-8-21	19-10-22	498, 70
(9) Alagoas	Garça Torta	Andrade Júnior	15-9-22	7-7-22	152, 52
(10) São Paulo	Santa Maria	Júlio Põrto	15-4-22	12-8-22	211, 55
(11) Bahia	Maraú	Paulino Carvalho	26-6-22	19-11-23	387, 30
(12) São Paulo	Itirapina, Rio Clarc	Gerson Alvim	17-7-22	31-10-23	324, 54
(13) São Paulo	Santa Maria, São Pedro	Júlio Põrto	4-10-22	24-9-23	251, 24
(14) Alagoas	Riacho Doce	J. F. Andrade	13-12-22	12-2-23	41, 95
(15) São Paulo	São Pedro	Bourdot Dutra	29-2-23	26-10-24	477, 53
(16) Paraná	Marechal Mallet	Paulino Carvalho	28-7-23	5-4-24	283, 72
(17) Alagoas	Riacho Doce	Eufrásio Borges	15-6-23	3-6-24	220, 50
(18) São Paulo	Tucum, São Pedro	Júlio Põrto	28-11-23	30-4-24	147, 11
(19) São Paulo	Itirapina, Rio Claro	Gerson Alvim	18-1-24	30-5-25	100, 05
(20) Bahia	Maraú	Afonso Galeão	25-9-24	23-7-25	240, 14
(21) Alagoas	Riacho Doce	Eufrãsio Borges	11-11-24	25-10-27	45–35
(22) Paraná	Marechal Mallet	Paulino Carvalho	6-12-24	7-3-27	518, 00
(23) São Paulo	Araquá, São Pedro	Bulhões Pedreira	22-1-25	6-8-28	380, 67
(24) Pará	Itaiúba	Pedro Moura	20-7-25	23-9-26	445, 10
(25) São Paulo	Graminha, São Pedro	Olinto Pereira	31-7-25	19-11-27	469, 01

(continued)

D. Peyerl, *The Oil of Brazil*, Historical Geography and Geosciences,
https://doi.org/10.1007/978-3-030-13884-4

Table B.1 (continued)

SONDAGENS PARA PETROLEO EFETUADAS PELO GOVÈRNO FEDERAL EM TODO O TERRITORIO NACIONAL, de 1919 a 1928					
Estados	Localidades	Encarregados	Início	Fim	Prof. metros
(26) São Paulo	Alambari, Botucatu	Couto Fernandes	1-12-25	1-12-27	446, 02
(27) Bahia	Santo Amara	E. Scorza	15-1-26	20-10-26	91, 60
(28) Rio Grande do Sul	Bela Vista, São Gabriel	Axel Löfgren	8-2-26	25-4-26	240, 14
(29) Pará	Bom Jardim, Itaituba	Pedro Moura	16-11-26	24-4-26	201, 58

De modo geral, o tipo de sondas empregado foi o de "Ingersol Hand, Rotativa", funcionando com aço granulado. O alcanoe máximo, verificado, dèstes tipos de sonda, não passava de 600 a 800 metros de profundidad
Com èsse modèlo de sonda foi atingida, por mim, em Marechal Mallet, no Paraná. (3.* na relação supra) a pro fundidade máxima de 510 metros, a maior até então alcançada, quando se deu o desprendimento de gäs combustivel conforme anotamos anteriormente, tendo conseguido atra-vessar uina espêssa camada de diabásio, rocha excessivamęnte dura e diaclasa
Source Gonsalves (1963): 143

Appendix C
Drills

See Figs. C.1 and C.2.

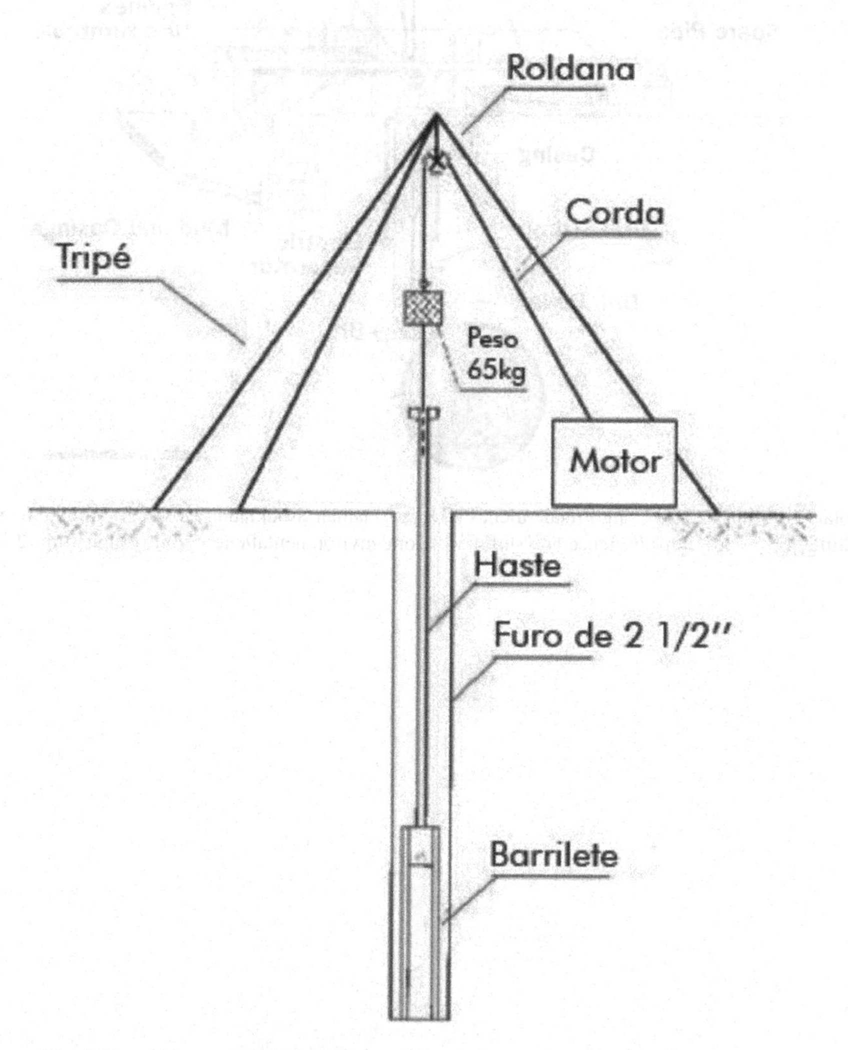

Fig. C.1 Percussion drilling. *Source* SONDAGEM A … s.d

D. Peyerl, *The Oil of Brazil*, Historical Geography and Geosciences,
https://doi.org/10.1007/978-3-030-13884-4

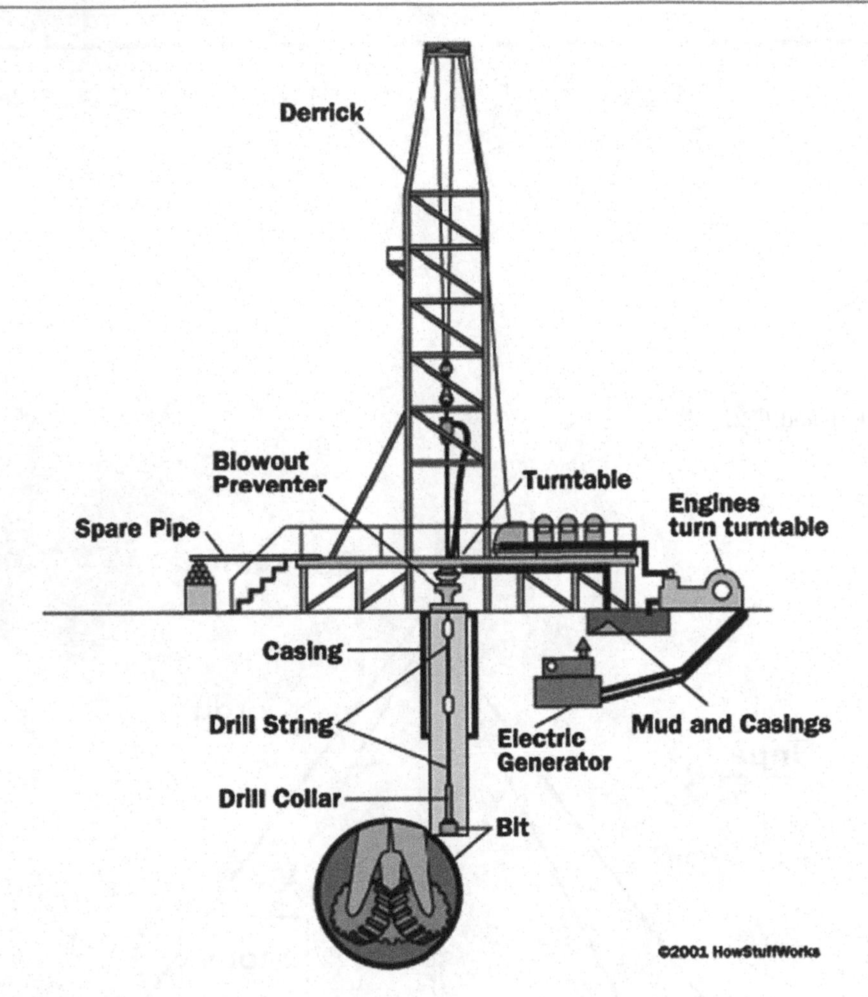

Fig. C.2 Rotary drilling. *Source* Craig Freudenrich, Ph.D. & Jonathan Strickland "How Oil Drilling Works" 12 April 2001. HowStuffWorks.com, https://science.howstuffworks.com/environmental/energy/oil-drilling.htm. 26 March 2019

Fourth Cover

The oil industry was born very late in Brazil—even when compared to other Latin American countries. The emergence of this productive activity occurred as a consequence of the persistence of public and private national actors who bet on the potential of the geological country. In this singular trajectory, it is also evidenced that the institutionalization of the geosciences occurred in the wake of the productive activity. It was when the Brazilian state committed itself to developing the country's oil activity, through the creation of the National Petroleum Council in 1938 and then of Petrobras in 1953, which revealed a lack of qualified human resources in the field of geosciences. This last aspect is particularly impressive in a territory of continental dimension and of so much mineral wealth. This unique experience is described in a captivating and masterful way by Drielli Peyerl. It takes up this history from the dawn of the oil industry in 1864, when the first decree was issued authorizing the mining of this mineral in the country until the end of the 1960s, when Petrobras made its first advances in offshore production. The Brazilian case clearly illustrates the thesis that progresses in the formation of human resources and, later, in research, only occurred as a result of an explicit development project. In fact, it was from the imperative of industrialization and the consequent expansion of the domestic demand for petroleum derivatives that the need arose to produce it on national soil. However, it was only with the expansion of its industry in the country that the need to train qualified human resources was consolidated, contributing to the sedimentation of geological science in Brazil. In this sense, Petrobras played a decisive role in the creation of the first geology courses in our context, and it is incumbent on it to act as a development agency in the institutionalization of this discipline in the country.

André Tosi Furtado

D. Peyerl, *The Oil of Brazil*, Historical Geography and Geosciences,
https://doi.org/10.1007/978-3-030-13884-4

Sources

A confusão do petróleo (1960) Jornal do Brasil, 2 Dec

A polêmica sobre o petróleo brasileiro (1961) Revista PETROBRÁS. Rio de Janeiro, ano 7, n 184, June

A questão do petróleo (1979) Política Mineral e Energética. Jornal do Geólogo, October/November/December

Amaral IC do (1961) Rio de Janeiro [To] Lange, Frederico Waldemar. Pedido de informação do deputado Pereira Silva. Archive Frederico Waldemar Lange, Box 113

Abreu SF Petróleo (1948) Boletim Geográfico, ano 6, n 62, May

Atingidas e ultrapassadas as metas do atual governo (1960). Jornal O Jornal, 2 Feb

Brasil. Decreto nº 393, de 12 de maio de 1890. http://legis.senado.leg.br/legislacao/listaPublicacoes.action?id=65464&tipodocumento=dEC&tipoTexto=PUB. Accessed Apr 2013.

Brasil. Decreto nº 3.352-A, de 30 de novembro de 1864. http://legis.senado.leg.br/legislacao/listaPublicacoes.action?id=102307&tipodocumento=dEC&tipoTexto=PUB. Accessed Apr 2013

Brasil. Decreto nº 5.014, de 17 de julho de 1872. http://legis.senado.leg.br/legislacao/listaPublicacoes.action?id=73823&tipodocumento=dEC&tipoTexto=PUB. Accessed Apr 2013

Brasil. Decreto nº 8.840, de 05 de janeiro de 1883. http://legis.senado.leg.br/legislacao/listaPublicacoes.action?id=68899&tipodocumento=dEC&tipoTexto=PUB. Accessed Apr 2013

Brasil. Decreto nº 21.079, de 24 de fevereiro de 1932. Diário Oficial da União, Brasília, dF, 26 mar. 1932. http://www.jusbrasil.com.br/diarios/2023923/pg-29-secao-1-diario-oficial-da-uniao-dou-de-26-03-1932/pdfView. Accessed May 2010

Brasil. Diário Oficial da União nº 448, de 26 de março de 1932. http://www.jusbrasil.com.br/diarios/2023923/pg-29-secao-1-diario-oficial-da-uniao-dou-de-26-03-1932/pdfView. Accessed May 2010

Brasil. Decreto n° 21.414, de 17 de maio de 1932. http://legis.senado.leg.br/legislacao/listaPublicacoes.action?id=33856&tipodocumento=dEC&tipoTexto=PUB. Accessed Jan 2013

Brasil. Decreto nº 21.415, de 17 de maio de 1932. http://legis.senado.leg.br/legislacao/listaPublicacoes.action?id=33863&tipodocumento=dEC&tipoTexto=PUB. Accessed Jan 2013

Brasil. Decreto n° 23.345, de 10 de novembro de 1933. http://legis.senado.leg.br/legislacao/listaPublicacoes.action?id=38156&tipodocumento=dEC&tipoTexto=PUB. Accessed Jan 2013

Brasil. Decreto nº 23.396, de 27 de fevereiro de 1934. http://legis.senado.leg.br/legislacao/listaPublicacoes.action?id=31070&tipodocumento=dEC&tipoTexto=PUB. Accessed Apr 2013

Brasil. Decreto n° 1.849, de 03 de agosto de 1937. http://legis.senado.leg.br/legislacao/listaPublicacoes.action?id=152305&tipodocumento=dEC&tipoTexto=PUB. Accessed Jan 2013

Brasil. Decreto-Lei nº 538, de 07 de julho de 1938. http://legis.senado.leg.br/legislacao/listaPublicacoes.action?id=103275&tipodocumento=dEl&tipoTexto=PUB. Accessed Jan 2013

Brasil. Decreto-Lei n° 1.143, de 09 de março de 1939. http://legis.senado.leg.br/legislacao/listaPublicacoes.action?id=10299&tipodocumento=dEl&tipoTexto=PUB. Accessed Apr 2013

Brasil. Art. 153 da Constituição Federal de 1946. www.jusbrasil.com.br/topicos/10614743/artigo-153-consituicao-federal-de-18-setembro-de-1946. Accessed June 2013

Brasil. Decreto n° 29.006, de 20 de dezembro de 1950. http://legis.senado.leg.br/legislacao/listaPublicacoes.action?id=161095&tipodocumento=dEC&tipoTexto=PUB. Accessed May 2013

Brasil. Lei n° 2.004, de 03 de outubro de 1953. www.planalto.gov.br/ccivil_03/leis/L2004.htm. Accessed Jan 2013

Brasil. Decreto nº 40.783, de 18 de janeiro de 1957. http://legis.senado.leg.br/legislacao/listaPublicacoes.action?id=171995&tipodocumento=dEC&tipoTexto=PUB. Accessed Jan 2013

D. Peyerl, *The Oil of Brazil*, Historical Geography and Geosciences,
https://doi.org/10.1007/978-3-030-13884-4

CENTRO DE APERFEIÇOAMENTO E PESQUISA DA PETROBRAS – CENAP. Normas reguladoras do Curso de Revisão. 1957a. CEnPEs/Petrobras Library
CENTRO DE APERFEIÇOAMENTO E PESQUISA DA PETROBRAS – CENAP. Curso de Geologia – Programa Provisório para o período de 1° de abril a 30 de junho. [s. l.: s.n.], 1957b. Archive Frederico Waldemar Lange, Box 110
Contestam os autores que a Petrobrás se oriente por determinados técnicos (1960). Jornal O Globo, 1 Dec
Dias J de NT (1991) José de Nazaré Teixeira Dias: Interview [1988]. Rio de Janeiro, CPDOC/FGV – SERCOM/ Petrobrás, p 374
Em 1932, falar de petróleo no Brasil era "mistificação" (1974) depoimento Pedro Moura – 03. Jornal O Globo. Archive Frederico Waldemar Lange, Box 01
Encontro de geólogos (1966) Porto alegre. Anais.... Porto alegre: Universidade Federal do rio Grande do Sul/ Escola de Geologia. Sob patrocínio do Conselho Nacional de Pesquisas
Entrevista Viktor Leinz (1982) Jornal do Geólogo – a história da Geociências no Brasil através de depoimentos ao Jornal do Geólogo. São Paulo, sociedade Brasileira de Geologia
Esclarecimentos aos candidatos a Geólogos da Petrobrás (1965) Archive Frederico Waldemar Lange, Box 113, Divisão de Exploração – DIVEX/PETROBRAS, Rio de Janeiro, October
Exploração na Bahia. Archive Frederico Waldemar Lange, Box 115
Eugênio gudin quer mandar muito mais do que café! Jornal Última Hora. Rio de Janeiro, 15 de dezembro de 1954
Famoso geólogo francês visita o Brasil (1959) Revista PETROBRÁS, Rio de Janeiro, ano 6, n 153, 1 Dec
Federação das Indústrias do Estado de São Paulo (1941) As industrias e as pesquisas tecnológicas. São Paulo, FIESP
Fortes AP (2003) CENAP – Petrobras: uma breve memória 1954–1964. Petrobras, CEnPEs/Petrobras Library
Frederico Waldemar Lange em trabalho de campo pela Petrobras. Meados da década de 50. Archive Frederico Waldemar Lange, Box 82
Gonsalves AD (1963) O Petróleo no Brasil: anotações do geólogo Alpheu Diniz Gonsalves – autobiografia de 58 anos de função como geólogo, 1905 – 1962. CENPES/Petrobras Library, Rio de Janeiro, Editora e Gráfica Polar
Gorceix CH (2011) Prefácio do annaes da Escola de Minas de ouro Preto n° 1, em 1881. REM – Revista Escola de Minas, v 64, n 3
Gubler YG (1967) Missão de assistência técnica junto à PETROBRÁS. Programa do Centro de Pesquisas (CEnPEs) da PETROBRÁS nos domínios da exploração e da produção. Archive Frederico Waldemar Lange, Box 114, Relatório. Rio de Janeiro
Humphrey FL (1961) Programa futuro proposto para o curso de Geologia de Petróleo. Archive Frederico Waldemar Lange, Box 114, Salvador [s.n.]
Interesse dos jovens brasileiros pela indústria nacional do petróleo (1959) Revista PETROBRÁS. Rio de Janeiro, ano 6, n 153, 1 Dec
Lange FW (1961) Aspectos Econômicos da Exploração do Petróleo no Brasil. Instituto Brasileiro de Petróleo, Rio de Janeiro
Lange FW [Letter] (1961) Rio de Janeiro [para] Presidente. organização regional e situação dos técnicosdo departamento de Exploração. Archive Frederico Waldemar Lange, Box 53
Lange FW [Note] (1961) Rio de Janeiro [To] Garcia, Evaldo da silva. visita ao Eni (Itália). nota sôbre as conversações realizadas no Rio de Janeiro, de 17 a 28 de julho de 1961, entre uma delegação do ente nazionale idrocarburi (eni), o CNP e a Petrobrás
Lange FW Pesquisa de Petróleo em Alagoas. Relatório interno da Petrobras. September 11 1961. Archive Frederico Waldemar Lange, Box 54
Lange FW [Letter] (1962) Ponta Grossa [To] Meijer. The possibility of employment with Petrobras. Archive Frederico Waldemar Lange, Box 79
Legislação Brasileira do Petróleo: 1947–1961. v 3. do CEnPEs/Petrobras Library
Link W [Letter] (1959) Rio de Janeiro [para] Lange, Frederico Waldemar. 1 f. Transferência. Archive Frederico Waldemar Lange, Box 114
Link W [Letter] (1960) Rio de Janeiro [To] Magalhães, Juracy. Bahia. 6 f. Archive Frederico Waldemar Lange, Box 108
Link W (1961) Exploração – PETROBRÁS – outubro 1954 até dezembro 1960. Report. Archive Frederico Waldemar Lange, Box 110
Ludwig G (1964) Nova Divisão Estratigráfica e Correlação Faciológica por meio de pequenas estruturas internas dos sedimentos silurianos e devonianos na Bacia do Médio Amazonas. Archive Frederico Waldemar Lange, Box 118, Centro de Aperfeiçoamento e Pesquisa de Petróleo – CEnaP; departamento de Exploração, Rio de Janeiro
Mattoso S (2012) O curso de Geologia da Petrobras. Interview by Drielli Peyerl
Maya E de (1938) O Brasil e o drama do Petroleo. Rio de Janeiro, José Olympio
Mentira velha: Brasil sem petróleo ...(1961) Jornal O Semanário, Paraná, year 6, n 244, 7 and 14 Jan
Ministério da agricultura (1920) Jornal O Paiz, Rio de Janeiro, 27 Apr
Moggi AS [Carta] (1957) [To] Sr. Geólogo-Chefe do DEPEX. Lecture Prof. Lange. 1 f
Moggi AS [Letter] (1961) Rio de Janeiro [To] Aguiar, Manoel Pinto de aguiar. 2 f. Archive Frederico Waldemar Lange, Box 114
Moggi AS [Letter] (1961) Rio de Janeiro [To] Aguiar, Manoel Pinto de Aguiar. Diretrizes CENAP. 9 f. Archive Frederico Waldemar Lange, Box 114
Moggi AS (1968) Pessoal para o avanço tecnológico – a experiência da PETROBRÁS. Jornal Diário de Notícias, Archive Frederico Waldemar Lange, Box 114, Rio de Janeiro, January

Moggi AS (1988) Antonio Seabra Moggi: depoimento. Interbviewer: José Luciano de Mattos Dias e Margareth Guimarães Martins. Rio de Janeiro, Petrobrás; Serviço de Comunicação Social; Memória da Petrobrás; Fundação Getúlio Vargas; Centro de Pesquisa e documentação de História Contemporânea do Brasil

Oliveira C (1961a) Resumo Histórico do Treinamento na Petrobrás. Boletim Técnico PETROBRÁS, Rio de Janeiro, v 4, n 1–2, pp 71–72, January/June

Oliveira C (1961b) Resumo Histórico do Treinamento na PETROBRÁS II Boletim Técnico PETROBRÁS, Rio de Janeiro, v 4, n 3–4, pp 141–144, July/December

Oliveira C (1962) Resumo Histórico do Treinamento na PETROBRÁS III Boletim Técnico PETROBRÁS, Rio de Janeiro, v 5, n 3, pp 71–72, July/September

Oliveira EP de (1940) História da Pesquisa de Petróleo no Brasil. Serviço de Publicidade agrícola, Rio de Janeiro

Oliveira Jnior EL de (1959) Ensino Técnico e Desenvolvimento. Ministério da Educação e Cultura, Instituto Superior de Estudos Brasileiros, Rio de Janeiro

Organograma básico da Petrobrás (1955) CEnPEs/ Petrobras Library

Ouro Negro no Brasil (1939) Jornal O Globo, 24 Jan

Passarinho JG [Letter] (1958) Belém [To] LinK, Walter. Relatório da viagem ao rio Moa. 3 f. Archive Frederico Waldemar Lange, Box 108

Pedreira A de B (1927) A pesquiza de petroleo. Typographia do «Annuario do Brasil», Rio de Janeiro

Perguntas e respostas (1959) Revista PETROBRÁS. ano 6, n 153, Rio de Janeiro, 1 Dec

Petrobras vai contratar técnicos francêses para ver se petróleo existe (1961) Jornal Diário de Notícias. 36, n 252, Porto Alegre, 1 Jan

Petrobras (1957a) A PETROBRÁS prepara o seu pessoal técnico: manual. Archive Frederico Waldemar Lange, Box 114, Centro de Aperfeiçoamento e Pesquisa de Petróleo, Rio de Janeiro

Petrobras (1957b) Programa Provisório do Curso de Introdução a Geologia. [s. l.: s.n.]. Archive Frederico Waldemar Lange, Box 110

Petrobras (1957c) Formação e Aperfeiçoamento de pessoal. Report. Archive Frederico Waldemar Lange, Box 114, Diretoria Executiva da Petróleo Brasileiro S.A PETROBRÁS,Rio de Janeiro

Petrobras (1957d) Resolução n° 7/57. Formação e aperfeiçoamento de Pessoal. In: Report. Archive Frederico Waldemar Lange, Box 114, Rio de Janeiro [s.n.]

Petrobras (1959a) Centro de aperfeiçoamento e Pesquisa de Petróleo. Curso de Manutenção de equipamentos de petróleo: manual. Archive Frederico Waldemar Lange, Box 114, Rio de Janeiro [s.n.]

Petrobras (1959b) Centro de aperfeiçoamento e Pesquisa de Petróleo. Curso de Refinação de Petróleo. Archive Frederico Waldemar Lange, Box 114, Rio de Janeiro [s.n.], pp 7–9

Petrobras (1960) Relatório preliminar do Grupo de Trabalho instituído pela Resolução n° 25/60, da Diretoria Executiva, para estudo da criação de um órgão de pesquisas para a indústria do petróleo. Archive Frederico Waldemar Lange, Box 114, Rio de Janeiro [s.n.]

Petrobras (1963) Centro de aperfeiçoamento e Pesquisa de Petróleo. Curso de engenharia de petróleo. CENPES/Petrobras Library, Rio de Janeiro [s.n.]

Petrobras (1966) Manual de Geologia de Superfície. N. 2. Archive Frederico Waldemar Lange, Box 14, Centro de Pesquisas e Desenvolvimento (CENPES), Departamento Industrial (DEPIN), Rio de Janeiro

Petrobras (1967a) Centro de Pesquisa e desenvolvimento. dez anos de Evolução Tecnológica. Boletim Técnico da PETROBRÁS, Archive Frederico Waldemar Lange, Box 113, Rio de Janeiro, v 10, n 1, January/ March

Petrobras (1967b) Relatórios Mensais. Archive Frederico Waldemar Lange, Box 32

Petróleo (1855) O Auxiliador da Industria Nacional. N 4, October

Pinto ID (2011) Interview - Irajá Damiani Pinto. By Drielli Peyerl. Porto alegre: [s.n.], Aug 2011

Pinto M da S (1988) Mário da Silva Pinto: depoimento [1987]. Interview. CPDOC/FGV – SERCOM/ Petrobrás, Rio de Janeiro, p 141

Pires do rio J (1937) O resultado do inquérito feito sobre o chamado caso do petróleo nacional: Um officio do Presidente da Comissão ao Ministro da agricultura. Archive Frederico Waldemar Lange, Box 32, Rio de Janeiro [s.n.], p 1

Primeiros geólogos diplomados no Brasil. Jornal do Estado de Minas, 1960. http://www.mmm.org.br/ media/usuarios/511/imagens/geologos3.jpg. Accessed 14 May 2013

Produção mundial de petróleo bruto – 1955/1960. Revista Petrobras. Archive Frederico Waldemar Lange, Box 32, Rio de Janeiro, ano 7, n 179, 1961, p 1

Recrutamento de técnicos para Landulpho Alves (1961) Revista PETROBRÁS, Rio de Janeiro, ano 7, n 184, June

Recrutamento e seleção de candidatos aos cursos do CENAP (1957) Boletim Técnico da PETROBRÁS, Rio de Janeiro, ano 1, n 1, October

Reunião de estudos técnicos de petróleo, between 5 and 11 October, 1958. Archive Frederico Waldemar Lange, Box 115, Bahia, Petrobras, 1958

Seabra O da S (1965) A indústria petroquímica no Brasil. Boletim técnico PETROBRÁS, Archive Frederico Waldemar Lange, Box 32., Rio de Janeiro, v 8, n 1, pp 115–133, January/March

Um Grande empreendimento econômico lançado no Brasil (1951) Jornal do Brasil, Rio de Janeiro, 5 Dec

Universidade Federal de Ouro Preto – UFOP. Escola de Minas de Ouro Preto. Relação dos Formandos de 1878 a 2007. http://www.semopbh.com.br/arquivos_pdf/ livro.pdf. Accessed Aug 2012

Walter Link deixando o Brasil: 'Cumpri o meu dever' (1961) Revista O Cruzeiro, Rio de Janeiro, ano 21, n 1

Williams FC et al (1960) Boletim Técnico da Petrobrás. Archive Frederico Waldemar Lange, Box 48, CENAP, v 3, n 2, Rio de Janeiro, pp 161–166, April/June

References

Costa DFO da [s.d.] Dos antecedentes da descoberta de Candeias ao Relatório Moura – Oddone, um passeio documentado por um período importante da história da exploração de petróleo no Brasil [s.l.]. Petrobras

Diógenes de Almeida Campos: pedras e ossos do ofício. http://revistapesquisa.fapesp.br/2005/02/01/pedras-e-ossos-do-oficio/. Accessed July 2011

Ensaios cronológicos sobre os precursores da Geologia no país. http://www.cprm.gov.br/publique/cgi/cgilua.exe/sys/start.htm?infoid=517&sid=8. Accessed Apr 2012

Futai MM (2013) História da Mecânica dos Solos no Brasil. http://www.futai.com.br/page14.php. Accessed Apr 2013

Iannuzzi R, Frantz JC (2007) 50 anos de geologia: irajá damiani Pinto: história e memória. Editora Comunicação e identidade, Porto alegre

Mello JMC de, Novais FA (1998) Capitalismo tardio e sociabilidade moderna. In: Schwarcz LM (org) História da vida privada no Brasil: contrastes da intimidade contemporânea. Companhia das letras, São Paulo, pp 586–588

Nacional dos Centros de Pós-Graduação em Economia—ANPEC, pp 1–21.

O Petróleo no Brasil. http://blog.planalto.gov.br/wp-content/uploads/timeline/fotos/1939-1941.jpg. Accessed Apr 2013

Odell PR (1996) Geografia econômica do petróleo. Zahar, Tradução Jairo José Farias Rio de Janeiro

Perfuração. http://www.cprm.gov.br/publique/cgi/cgilua.exe/sys/start.htm?infoid=631&sid=23. Accessed Apr 2013

Petrobras 50 anos. http://cpdoc.fgv.br/producao/dossies/Fatosimagens/Petrobras50anos. Accessed Apr 2013

Sacchetta V (2012) Petróleo ainda que tarde. http://lobato.globo.com/novidades/novidades23.asp. Accessed Oct 2012

Sonda rotativa. Disponível em: http://ciencia.hsw.uol.com.br/perfuracao-de-petroleo2.htm. acesso em: 25 fev. 2013

Sondagem a percussão. http://www.nfsondas.com.br/sondagem_percussao.html. Accessed Feb 2013

D. Peyerl, *The Oil of Brazil*, Historical Geography and Geosciences,
https://doi.org/10.1007/978-3-030-13884-4